W9-ARG-263

A. Anikin

INTERNATIONAL
PUBLISHERS

Translated by K. M. Cook
Designed by Yu. Lyuter

First United States edition 1983 by International Publishers

Printed in the Union of Soviet Socialist Republics

Library of Congress Cataloging in Publication Data

Anikin, Andrei Vladimirovich.
Gold—the yellow devil.

Translation of: Zheltyi d′iavol.
Includes bibliographical references.
1. Gold. I. Title.
HG289.A6413 1983 332.4'222 83-274
ISBN O–7178–0599–9

CONTENTS

Page

I. Introduction. Gold in the Life of Mankind 5

II. How Gold Became Money 20

III. Mining the Yellow Metal 36

IV. Where the Gold Goes 64

V. Gold Genocide 104

VI. The Power of Gold 122

VII. The Gold Standard and Its Agony 144

VIII. The International Monetary System 170

IX. The Fate of the Yellow Metal 202

Notes 237

I. INTRODUCTION.
GOLD IN THE LIFE OF MANKIND

This book was first published in Russian at the end of 1978. Subsequent events have confirmed its relevance today. Many a time the newspapers of the capitalist world have carried headlines screaming "Gold Panic!", "Gold Rush!" In graphs of gold prices the jagged line on the right suddenly soared up almost at right angles. It soon became impossible to chart them on the same scale, because at the beginning of 1980 the market price of gold in London or New York was almost 25 times higher than it had been only ten years earlier.

Then, in 1980-1982, the price of gold fell by 50 per cent and was on average about 400 dollars an ounce in the first half of 1982. The main reason was that it was a period of prolonged economic recession, when demand for most commodities dropped and in many capitalist countries there was a reduction in production and capital investment. These sharp fluctuations in the price of gold both reflect and enhance the instability of the capitalist economy.

Up to 1971 the official price of gold was 35 dollars per troy ounce (1.1 dollar per gramme), but it could be obtained legally from the US Treasury at this price only by a single privileged category of buyers—foreign monetary authorities and central banks. At that time this price differed only slightly from the free market price at which it was purchased by jewelers, manufacturers of electronic equipment, and rich investors who thought it prudent to invest part of their capital in gold. Buyers of jewelry and gold coins paid more per ounce of gold, of course: they were covering the cost of manufacturing the jewelry and coins, the profit of the manufacturers and retailers, and various taxes and surcharges. But the difference was still not very great, prices being in the order of a few dozen dollars per ounce.

At the height of the gold rush in mid-January 1980 the wholesale price went up to 840-850 dollars per ounce, approximately 27 dollars a gramme of pure gold. Jewelers did not have time to change the price labels in their glittering shop windows. Demand was vastly in excess of supply.

Gold was not alone in this stock jobbing. The prices of silver, platinum and metals of the platinum group also shot up astronomically. And in the background, influenced by other forces and subject to other laws, the price of oil rose steadily. Such was the situation in which capitalism entered the 1980s.

There are more important things in the world than the gold rush, of course. To take an example from the economic sphere, the increase or fall in oil prices is more significant and of far greater consequence.

Nevertheless, the economic (and political!) importance of gold is far greater than one would think from its fairly modest and limited spheres of industrial use. It is also incomparably greater than that of the physico-chemically related metals—silver and platinum.

Commodities and Money

Gold occupies a very special place in the vast world of modern commodities. This place is determined by its specific social functions. It has not only a lustrous appearance, but also an illustrious history, without which we cannot understand its present-day role.

Socialism inherited gold as an economic phenomenon from capitalism. Insofar as under socialism there is buying and selling of commodities and the functioning of money, gold retains its role here as a highly specific commodity with special economic and socio-psychological properties. They differ greatly from the features and functions of gold in capitalist society. This can be seen if only from the fact that the recent gold rush, detrimental from the economic and distasteful from the moral point of view, did not and could not have an analogy in socialist countries. We do not possess the social stratum of private dealers, speculators and stock-exchange gamblers who continue to make money on the gold rush at the expense of those who are inexperienced, uninformed and often by no means wealthy. But gold is bought and sold in socialist countries, and many people seek to possess it for a variety of reasons. Even if we accept that these reasons are irrational and not in keeping with the essence of the socialist way of life, we must take them into account as an economic and social reality.

Let us take a closer look at this metal that has always attracted special attention and interest. Everyone must have seen gold. Many readers will possess objects made of gold: rings, bracelets, watches, or gold crowns. Many of them will probably have seen a gold coin. At one time these circulated as money, but now they are kept as precious objects. Sometimes they are of historical and artistic value and form part of numismatic collections.

We know that gold is a very expensive commodity. More precisely, that it is highly valuable in small amounts. Any weight unit of silver costs tens of times less, copper thousands of times less, and iron tens of thousands of times less than gold. As we know, gold is called a *precious* metal.

Everyone will agree that gold is beautiful. It has an intense yellow which can take on different shades depending on the quantity and type of admixtures (usually silver, copper, nickel and certain other less valuable metals). The very word *gold* in Indo-European languages is derived from the word for yellow. In particular, this concerns the Slavonic and Germanic languages, including English (although the relationship is more obvious in German: *Gold* and *gelb*). From time immemorial people have made all sorts of ornaments and jewelry of gold. Sometimes these are of great artistic value and important cultural and historical monuments. The world's museums treasure Ancient Egyptian, Scythian and Thracian gold articles, and the work of skilled mediaeval and modern craftsmen. The modern jewelry industry produces many articles of the most varied artistic and monetary value.

Gold possesses some unique physical and chemical properties. It does not rust like iron, does not get covered with a bluish-green oxide like copper, and does not even darken like silver. Gold coins and other articles found in the ground or under water often look as if they have just come out of the workshop where they were made. This is why gold is called a *noble* metal.

The natural properties of gold make it suitable for use in

certain spheres of industry. It has acquired this function mainly in recent decades: it is used in electronics, space technology and communications.

But the importance of gold in world history and in modern economics is determined not by its use in jewelry manufacture or in industry. Moreover, we shall see that its ornamental use is connected not only and not so much with its useful natural properties, as with its social function—to embody value, to be *money.*

The role played by money in society depends on the structure of society. In a certain sense the reverse is also true: when we see how money functions in a certain society, we can judge its nature, the types of property, the character of economic relations between people.

The ages of human development differed from one another, inter alia, in the type of money they had. Homeric Greece did not yet have metal coins, and pieces of iron and copper or cattle served as money. In the last century gold and silver coins circulated throughout the world, and also banknotes which in normal conditions could be exchanged for metal at their face value. Today in the capitalist countries money for the most part takes the form of entries in bank accounts and is used for bank transfers. Ready money (banknotes and coins of non-precious metals) forms the smaller part of the money supply. In the USSR money takes the same forms, but the sphere of its operation is limited and subordinated to the interests of society: no one can use money to buy land, a share in an enterprise or a block of flats. Socialist enterprises settle their accounts by bank transfer exclusively, by means of transferring money from one account in the State Bank to another. Payment by bank transfer between members of the public and state enterprises is being introduced. The vast majority of payments made by members of the public is done in ready money (banknotes of the State Bank, treasury notes and small change). The planning and regulating of money supply is an important part of socialist economic planning.

It would be correct to call all these types of money, both in the capitalist and in the socialist economy, *paper-credit money.* Unlike the gold or silver money which used to be in circulation, it has no intrinsic value. The purchasing power of a 100-dollar or 100-rouble banknote bears no relation to the cost of its manufacture. This money is of a *credit* nature in the broad sense of the word, insofar as it is based on trust in the body that issues it. And the word *credit* in Latin does actually mean trust. We have grown accustomed to call this money paper money, although it is becoming more and more "dematerialised", taking the form of entries in bank accounts and, more recently, of electronic signals in the memory of computers.

Is gold money in a socialist economy? Economists give different answers to this question, some affirmative, others negative. But they all tend to qualify the answer with various reservations and specifications.

In fact it is by no means obvious from day-to-day experience. On the one hand, we are accustomed to think of gold as money. It is usual to think that notes can circulate only because they are in some way backed by gold.

On the other hand, you cannot go into a shop with gold in your purse, nor can you use it to pay the rent. It never occurs to anyone to say that a car or a television costs so many grammes of gold. All goods in the USSR have prices in Soviet roubles, and in this respect gold does not differ from any other commodity. If a person has some gold and he needs money—ordinary paper money, he must first of all sell his gold. Take it to a second-hand shop, or pawn it, if the owner cherishes his grandfather's watch or grandmother's bracelet as a memento. But a fur coat, or a transistor radio, or silver spoons can also be sold or pawned. Is there any difference? Obviously there is. Gold differs in that it is highly portable, homogeneous and long-lasting. Moreover, it possesses another quality, hard to define but real nevertheless: that of age-old tradition, the custom of distinguishing it from all other commodities. When people want to ascribe a special value to something verbally in Russian, they say "white gold" for cotton or "black gold" for oil.

Many Soviet economists studying money in socialist society consider that gold has practically lost all monetary functions in the USSR. The purchasing power and stability of paper-credit money is based not on gold, but on the supply of commodities that are produced and put into circulation, for the most part at state-fixed stable prices.

The state's gold reserves play a certain role in maintaining the stability of paper-credit money, because gold is in itself a very valuable commodity which is easily realised abroad. In the Great Soviet Encyclopaedia we read: "In socialist countries the central gold reserves serve mainly as a reserve for international settlements connected with external economic relations."[1] The Soviet rouble has a gold content that is fixed by law. On 1 January 1961 it was set at 0.987412 grammes of pure gold. Previously it had been equal to 0.222168 grammes since 1 March 1950. The increase was made in connection with a change in the scale of prices, when all monetary units in the national economy were recalculated at a ratio of 10:1 (lowered ten times) and new banknotes were issued. The gold content of the rouble, like the gold reserve, plays a real part primarily in the sphere of external economic relations. It serves as a base for fixing the ratio of the rouble to foreign currencies (exchange rates). The last fixed ratio (parity) of the rouble and the dollar, based on the gold content of both currencies, was (from February 1973) 74 roubles 61 kopecks to 100 dollars, that is, 1 dollar was approximately equal to 75 kopecks. However, since then the gold content of the dollar has lost all real economic significance, and its rate in relation to other currencies has constantly fluctuated on the market. Therefore the afore-mentioned parity of the rouble and the dollar is used now only as a point of reference, and the rate of the rouble in relation to the dollar is fixed by the State Bank of the USSR taking into account the market rate of the dollar in relation to a set of foreign currencies.

Other Soviet specialists consider that in the USSR paper-credit money is connected with gold, not only because of the function of gold as world money, but also because of its role in purely internal economic processes. They believe that paper-credit money obtains its purchasing power in relation to commodities only in its capacity as a representative of gold.

The reader will probably already have guessed that the author of this volume is closer to the former viewpoint.

But whatever the divergencies of opinion on the question of the relationship between gold and paper-credit money, everyone is agreed that gold plays an important role in the foreign economic sphere, particularly in the relations of the USSR with capitalist countries. This role derives from the fact that the USSR is a major gold-producing country. As early as 1921 Lenin expressed most vividly the attitude of Communists to gold in his article "The Importance of Gold Now and After the Complete Victory of Socialism". The article was written during the introduction of the New Economic Policy, when it was important to re-establish normal commodity and monetary relations in the country ravaged by many years of war and establish trading relations with the capitalist West. Under these conditions the question of gold acquired considerable importance. Lenin wrote: "When we are victorious on a world scale I think we shall use gold for the purpose of building public lavatories in the streets of some of the largest cities of the world. This would be the most 'just' and most educational way of utilising gold for the benefit of those generations which have not forgotten how, for the sake of gold, ten million men were killed and thirty million maimed in the 'great war for freedom', the war of 1914-18... Meanwhile, we must save the gold in the R.S.F.S.R., sell it at the highest price, buy goods with it at the lowest price."[2]

Gold is a symbol of bourgeois barbarity and, in a certain sense, a synonym for capital, which engenders world-wide disasters. But because it is greatly prized in capitalist countries, we can and should use it in the interests of building a socialist economy. Lenin's practical advice sounds particularly relevant today, when the price of gold on the world market is rising in leaps and bounds, sometimes by as much as 40 or 50 dollars an ounce in a day. In 1980 the gap between the lowest and the highest market price was as much as 400 dollars an ounce. In this state of affairs the seller of gold has ample opportunity to show his commercial know-how, foresight and restraint.

What awaits gold in the more distant future, when the form of organisation of society changes on a world-wide scale?

Most probably future generations will themselves decide how best to use the gold accumulated by mankind and what amounts of it to produce. One can hardly doubt that the monetary functions of gold will be completely abolished. The use of the metal will become rational and functional. By then the mark of the Devil will be erased from gold and it will cease to be a symbol of greed, ill-gotten wealth and the exploitation of man by man. First and foremost, it will obviously be used in technology. But its beauty will also serve to satisfy man's healthy aesthetic requirements.

A Little History

But let us turn from the future to the past. The title of this book was inspired by Maxim Gorky's pamphlet "The City of the Yellow Devil" in which he portrayed New York at the beginning of the 20th century as a monstrous product of a civilisation based on money. This image, which attains apocalyptic force, was a reflection of the real fact that at that time gold was the dominant form of money, in essence, a synonym for money. And money both in the ordinary sense and in scientific analysis is the initial form of capital.

The history of gold goes back over the millennia. Gold as money, as concentrated economic power, as an object of desire and greed appears in ancient times. The Roman poet Virgil first coined the phrase *auri sacra fames*—"the cursed thirst for gold". But it took capitalism to bring this formula to completion. Why? Because in capitalist society gold is not only money, but also capital, that is, money put into circulation to make it grow. Balzac's philosophising usurer Gobseck says: "What is life if not a machine which money sets in motion . . . Gold is the spiritual essence of all present societies."[3]

Gold as a metal and a useful natural raw material has been known to man for about six thousand years. It acquired certain monetary features in the most developed parts of the ancient world about 4,000-4,500 years ago. It is now 2,500-2,600 years since the first gold coins appeared. This marked the beginning of its life as money.

Terrible crimes have been committed for love of gold. "People perish for metal. . ." is a phrase that describes its history very accurately. The power of gold makes people cruel and morally corrupt. Naturally, the blame for this must be put not on the metal, which is simply a chemical element with certain properties, but on the social system which has made it the embodiment of wealth, an object of almost religious worship.

For many centuries gold's rival for the role of money was silver. Not until the 19th century did gold reign supreme as the money metal. A monetary system called the gold standard was introduced in the main capitalist countries. The state minted and issued gold coins—English sovereigns, French Napoleons, and Russian imperials. All other forms of money were exchangeable for these coins. Gold became the basis, the hub of the monetary system. Paper-money inflation in such an economy was impossible. When commodity prices went up, it meant that for some reason these commodities had grown more expensive in relation to gold.

The gold standard was established first and foremost in Great Britain, and throughout the century gradually spread to many other countries. This is why Karl Marx based his analysis of capitalism in the middle and the latter half of the 19th century, in particular, on the fact that gold could be regarded as the only form of money. He says in *Capital*: "Throughout this work, I assume, for the sake of simplicity, gold as the money-commodity."[4]

Gold played an important role in the development of capitalism, in the formation of its economic mechanism, and in social relations. It is an important element of the whole of bourgeois civilisation.

The inflow of gold (and silver) into Western Europe after the great geographical discoveries of the 15th and 16th centuries was an important source of the original accumulation of capital. It destroyed the closed feudal economy and stimulated commerce and handicrafts.

In the mid-19th century the discovery of gold in California and Australia served as a spur to rapid economic development in Western Europe and North America. Gold promoted the economic drawing together of countries and territories and the formation of a world market. The inflow of gold formed the basis for the introduction of the gold standard and the development of the credit system.

At the end of the 19th century gold was discovered in southern Africa and this marked the beginning, one might say, of the modern phase in its long history. Today the Republic of South Africa produces about three-quarters of the capitalist world's gold. In recent years all the capitalist countries taken together have produced from 950 to 1,000 tons of gold annually, including about 700 tons in South Africa. These figures do not change much and, in spite of the fantastic rise in the price of gold, its production is not increasing. This unusual economic behaviour links gold with oil, the production of which has shown a tendency to drop in recent years.

The huge concentration of gold production in South Africa is most important for its future in the coming years and decades. The situation in the turbulent south of Africa is unstable. The victory of the people of Zimbabwe and the growth of the national liberation movement in South Africa and Namibia is increasingly weakening the position of the racist state and international monopoly capital in this country. On the other hand, the future of South Africa itself is to a large extent connected with the gold-mining industry, with the labour and struggle of hundreds of thousands of African miners.

In its time gold had a direct and strong influence on the economic development of the regions in which it was mined. The cities of San Francisco, Sidney and Johannesburg grew up "on gold".

The fate of the gold varied. Before World War I a considerable part was minted into coins which were then put into circulation. About the same amount went into the vaults of the central banks and treasuries and served to back credit and paper money. In both cases the gold took the form of money and was constantly being transferred from the reserves into circulation and back again. This was the period of the gold standard.

After the 1914-1918 war most capitalist countries stopped minting coins and restricted the exchange of paper money for gold. All forms of this exchange were finally abolished in the period of the world economic crisis of 1929-1933. Gold ceased to be money in the national economies of the capitalist countries. It was increasingly concentrated in state reserves and central banks and emerged only to settle international debts. There was only one path for freshly mined gold—"from pit to pit", that is, from the mines to the armoured vaults of banks and state reserves.

In the 1930s state reserves devoured not only all the gold being mined, but also a considerable part of private holdings. By 1945 (the end of World War II), two-thirds of the capitalist world's gold reserves (almost 30,000 tons) were concentrated in central reserves, and of this amount a little less than two-thirds (about 18,000 tons) was in the hands of the government of the United States. The concentration of gold in the United States continued in the early post-war years: by the end of 1949 the United States possessed 70 per cent (about 22,000 tons).[5]

In the post-war monetary system the American dollar became as it were a duplicate for gold, in so far as the US government guaranteed that other states could exchange their dollar assets (foreign currency accumulations) for gold at a fixed rate. "The dollar is as good as gold" remained a popular saying in the United States right up to the 1970s.

Today it sounds ridiculous. In 1968 the United States was forced to stop supporting the dollar price of gold on the free market, and in 1971 to stop exchanging dollars accumulated abroad for gold. Thus, the final link of capitalism's international monetary system with gold was broken.

The functioning of a monetary system linked with gold has proved to be objectively impossible for modern capitalism. At the same time the concrete forms and periods for ousting gold are determined largely by the policy pursued by the leading capitalist countries. In the 1970s the United States pursued the policy of the *demonetisation* of gold, i.e., of removing it from the monetary system as the basis for measuring the value of currencies and as the ultimate medium for international settlements.

In 1976 the member countries of the International Monetary Fund concluded an agreement, one of the most important aspects of which was the abolition of currency par values expressed in gold, the ceasing of operations on the basis of these par values and the liquidation of a considerable part of the Fund's gold reserves. However, this is still a long way from the actual removal of gold from the monetary system. The capitalist countries keep their metal reserves as the most reliable reserve of international payment media and sometimes put them into circulation, selling or pawning gold at market prices. Certain countries, particularly oil-producing ones, are increasing their reserves by buying gold on the market. Thus there is a certain movement of gold between the central banks and the governments of various countries. It is particularly symptomatic that when the Common Market countries set up the European monetary system in 1979 a place was found in it for gold.

The repeated rises in the price of gold have led to a rise in the real value of gold reserves and their ability to settle balance of payments deficits. In fact this means that gold is performing an important money function.

The enormous demand for gold as a medium for investing money capital is explained by the growing inflation in the capitalist world, economic and political instability and international crises. The end of these phenomena is not in sight and, consequently, most specialists are predicting in the long run price rises on the gold markets.

No matter how the functions of gold and the various aspects of its economic role change, this role remains a major one in the capitalist world.

Through the Eyes of Scholars and Writers

A vast amount has been written about gold. There are numerous works on the chemistry, physics, geology, extraction technology and use of gold. But perhaps even more has been written about its socio-economic and monetary aspects. As Marx recalled, the English politician Gladstone, speaking in a Parliamentary debate in the 1840s, observed that even love had driven less men mad than meditation on money. And money at that time was gold.[6]

Subsequently academic works on money began to pay far less attention to gold, and frequently it has been ignored completely. This does not mean that economists have lost interest in it, but merely that there have been certain changes in the structure of economic theory. On the other hand, the number of writings devoted to gold in socio-economic literature is growing so rapidly that it is becoming hard to keep up with them. The complexity and variety of the subject compel authors to examine different aspects of the problem of gold—economic, social, psychological and others. Many of these works, both Soviet and foreign, have been used in the writing of this book. The reader will find the titles of these books and articles in the notes.

The folklore of all peoples and the fiction of all ages have preserved a lot of observations and thoughts on the role of gold in people's lives. We find gold lauded and cursed, we find reverence for and horror of its power, and dreams of a world without this power. Homer and Sophocles, Chaucer and Shakespeare, Pushkin and Gorky... The Californian gold rush gave us the great writer Bret Harte, the Alaskan—Jack London. The world of Russian gold magnates and prospectors is portrayed vividly by Mamin-Sibiriak and Shishkov. Mikhail Zoshchenko in *The Black Prince* writes about the search for an English ship laden with gold that sank in the Black Sea. The English writer Alan Sillitoe in his novel *A Start in Life* describes the gold business, the world of smugglers and their bosses. And so on.

Here is one example taken from English poetry. The talented satirical poet Thomas Hood (1798-1845) "lauds" gold in these dynamic lines:

Gold! Gold! Gold! Gold!
Bright and yellow, hard and cold,
Molten, graven, hammered, and rolled;
Heavy to get, and light to hold;
Hoarded, bartered, bought, and sold,
Stolen, borrowed, squandered, doled;
Spurned by the young, but hugged by the old
To the very verge of the churchyard mould;
Price of many a crime untold;
Gold! Gold! Gold! Gold!
Good or bad a thousand-fold!
How widely its agencies vary—
To save—to ruin—to curse—to bless—
As even its minted coins express,
Now stamped with the image of Good Queen Bess,
And now of a Bloody Mary![7]

Sometimes writers touching upon "gold problems" in their works of fiction consciously or unconsciously raise important socio-economic issues. The Soviet writer Alexei Tolstoy wrote a fantasy novel entitled *Engineer Garin's Hyperboloid* in the 1920s. The novel's main character, an adventurist by the name of Pyotr Garin, finds a way of extracting vast quantities of gold from the earth's crust with the help of apparatus which anticipated the modern laser. The cost price of this gold is 2.5 dollars a kilogram, whereas the official price of gold, which corresponded to the gold content of the American monetary unit at that time, was about 665 dollars a kilogram. By selling this gold at cost price on the US market, he undermines the country's economic system to such an extent that supreme authority falls into his hands like a ripe apple dropping into a basket.

Here is the appeal which the US government makes to the people in connection with Garin's actions: "The mine on Gold Island must be blocked up as soon as possible and the very possibility of having inexhaustible supplies of gold must be eliminated. What will happen to the equivalent of labour, happiness, life itself, if people start digging up gold like clay? Mankind will inevitably return to primitive times, to barter trade, to savagery and chaos. The whole economic system will collapse, and industry and commerce will die."[8] And so on in the same vein.

The interesting question for the economist is this: if Tolstoy's fantasy were transferred to the modern world, what consequences for the capitalist economy could one expect from a sudden drop in the price of gold to the price of copper or aluminium? At the end of the book we shall attempt to find an answer to this question.

In recent years profound changes have taken place in the status and role of gold in the capitalist economic system. Gold is, possibly, at a turning point in its long history. This book describes how it has arrived at the present stage of its history, what its economic role is today, and what awaits it in the future.

This book is designed for the general reader with an interest in economics, history and politics. No special knowledge is required in order to read it. The modern reader is intelligent and demanding. Therefore the book observes the accepted standards of scientific publications, especially references to the sources of facts and ideas of other writers. This is particularly important for readers who wish to extend the range of their reading and knowledge. But all information on the essence of the subject is contained in the main text. The book may be read without paying any attention to the source references.

The present edition is based on the published Russian text. It has, however, been considerably revised and supplemented. In so doing the author was guided by three considerations: first, to include material relating to 1978-1982 which has contributed much that is new to the position of gold in the capitalist system; secondly, to improve the exposition and fill in gaps; and, thirdly, to a certain extent to take into account the interests of the foreign reader.

In this book the author does not claim to have made an independent study of the problem. His aim was a different one: to set out in popular and, if possible, entertaining form, with the use of historical and factual material, the views and ideas developed by Soviet economists. Some of these works are referred to in the notes, others are not mentioned, but they too have played their part in the preparation of the book. Over the last ten to fifteen years a number of serious works on international monetary problems have come out in the USSR, which have made a more or less special study of the question of gold also. Some of them have been translated into foreign languages. Unfortunately the last work of fundamental importance devoted specially to this question was published a comparatively long time ago—in 1968. It is S. M. Borisov's book *Gold in the Economy of Modern Capitalism* (Finansy, Moscow). Three other books by Soviet authors deserve special mention: O. S. Bogdanov, *The Monetary System of Modern Capitalism (Main Trends and Contradictions)*, Mysl, Moscow, 1976; D. V. Smyslov, *The Crisis of the Modern Capitalist Monetary System and Bourgeois Political Economy*, Nauka, Moscow, 1979; and G. G. Matyukhin, *Problems of Credit Currency Under Capitalism,* Nauka, Moscow, 1977. They give a profound analysis of the development of the capitalist monetary system after World War II. The author shares many of the interesting ideas on the nature of money under modern capitalism which V. M. Usoskin develops in a number of his recent works.

At the same time the author is not always in agreement with the scholars mentioned and in certain cases develops his own point of view. Some of his views have been stated earlier in scientific articles. No one but the author is responsible for any possible errors or other shortcomings of the book.

II. HOW GOLD BECAME MONEY

So we are interested in gold as money. It is this that has made it "the most contradictory metal", a metal with unusual and complex social functions. In order to explain how and why gold became money, we must go back to antiquity. In many cases a model of antiquity is provided by the life of peoples who have survived up to modern times while retaining a primitive structure of society.

The Emergence of Money

In primitive communities, where people lived by gathering the fruits of the earth or hunting wild animals, what they obtained from these activities was simply divided out between the members of the community. There was no *barter* of the results of this primitive "production". There was no need for barter between communities and, consequently, there was no need for money either.

But then man began to tame animals and till the soil. Gathering and hunting gave way to cattle-breeding and farming. Gradually some tribes began to engage predominantly in the former, and others in the latter. In this situation barter became an essential element of human life and the development of production and society. It grew even more important when handicrafts developed as a separate sphere of human labour. At the same time changes were taking place in the primitive-communal system itself; elements of the future slave-owning society were gradually developing within it. More and more commodities were appearing and barter was becoming more regular. Whereas previously it had been a random business, and the relationship between the objects bartered not very definite, now each commodity sought to express its value in a multitude of other commodities and was in fact exchanged for them.

Let us imagine a potter who has taken his pots to the market hoping to exchange them for grain. Unfortunately for him all the grain-owner's pots are safe and sound and he has no need to exchange his grain for the potter's wares. If there were only these two producers at the market, they would probably have to go home taking their wares with them. But fortunately there are also sellers of wool, olive oil, bronze knives and lots of other things on the market place. And it is already accepted that a pot, for example, is valued at a pound of wool, a bottle of olive oil of a certain size or two bronze knives. The potter may have to make several acts of exchange before he obtains a commodity which satisfies the grain seller and thus he acquires eventually the grain he wants.

That all this time-consuming business is not pure imagination can be seen, for example, from the following description by a European who travelled around Tropical Africa in the middle of the 19th century: "The arrangement at the hiring [of a boat] was rather amusing. Syde's agent wished to be paid in ivory, of which I had none; but I found that Mohammed ibn Salib had ivory, and wanted cloth. Still, as I had no cloth, this did not assist me greatly until I heard that Mohammed ibn Gharib had cloth and wanted wire. This I fortunately possessed. So I gave Mohammed ibn Gharib the requisite amount in wire, upon which he handed over cloth to Mohammed ibn Salib, who in his turn gave Syde ibn Habib's agent the wished-for ivory. Then he allowed me to have the boat."[1]

The absence of a *universal equivalent,* a commodity which expressed the value of all other commodities and was capable of being exchanged for them constantly, was the first big difficulty of barter. By making barter difficult, it hampered the progress of social production. What we have here is a contradiction of social development that was seeking for its resolution and was to find it sooner or later. Where? In the selection from the whole varied commodity world of a single commodity that acquired the properties of a universal equivalent. The selection of such a commodity is not usually the result of agreement between people or of some sort of command. It is a spontaneous economic process that takes place irrespective of people's will.

When such a commodity is found, all the other commodities turn to face it and devour it with their eyes. They now express their value in it, which makes it possible to compare them with one another. They now seek to be exchanged for this commodity, in order to choose any necessary object from the commodity world without difficulty.

Now our potter could exchange his pots for this special commodity and then receive for it grain, olive oil and any other product. In fact this is not bartering but buying and selling; the potter sells pots in order to buy grain. If, in the part of Africa which the above-mentioned author is describing, a commodity, ivory, for example, had already acquired the status of a universal equivalent, his difficulties would have been solved far more easily: he would simply have exchanged his wire for ivory and paid for the boat with it.

We have not yet uttered the word *money*, but it is easy to see that a universal equivalent which measures all values and is constantly circulating on the commodity market is, essentially, money.

Reading books on social history and the history of money convinces us that it is hard to think of a commodity that has not performed the function of money at some time or other. Furs and sea-shells, fish and cocoa beans, cattle and salt, amber and even people themselves (slaves) have all been money.

In each case the commodity did not acquire the functions of money accidentally. This role can go to important commodities that the people of the community have acquired from outside, from other tribes and peoples. Captives or slaves are an example of such a commodity.

European travellers tell us that even at the end of the 19th century in certain remote parts of Africa, in the Western Sudan, the unit of value was a slave, and definite traditional exchange relations existed between this commodity and others of both local and European origin. The paradox was that West European silver coins were not money there, but one of the many commodities. They were acquired not in order to put them quickly into circulation again in return for another necessary commodity, but to keep them as ornaments for the home or one's person. If a native of one of these parts of Africa was asked the price of a horse, he might well have answered "three captives" (slaves). A bull could cost "half-a-captive". This unusual monetary unit was an old or sick slave, whereas a young, healthy slave was a full monetary unit. It is easy to call this barbaric, of course, but the peoples of Europe and America have passed through equally barbaric stages of development in their time.

However, this exotic barbarism is an exception in the history of money. As a rule monetary functions were acquired by a commodity which was an important object of production and served in a large quantity as an object of market exchange in the locality in question. In Iceland, where the people were sailors and fishermen, fish was money, naturally: the prices of commodities were reckoned in pieces of cod and all commodities were "bought and sold" for cod. In many countries, beginning in ancient times, cattle or sheep were money. In Homer's poems we frequently find various commodities valued in terms of bulls. Cattle in the function of money has played such an important role in history that this is reflected in the language of many peoples. The Latin word *pecunia* (money) comes from the word *pecus* (cattle). In English traces of this have remained in the word *pecuniary*. It is reflected differently in Russian. In Old Russia the prince's treasury was called the *skotnitsa* (the word for cattle being *skot*), and the treasurer the *skotnik,* although the treasury no longer contained bulls and cows, but furs and silver ingots. Furs also performed the functions of money, particularly in Old Russia and Scandinavia. Definite ratios were fixed between the furs of different animals (sable, marten, beaver, etc.).

One can imagine the difficulties people had when these commodities were used as money. Let us return to the example of our potter. Let us assume that he has exchanged his pots for a flock of twelve sheep. But the grain seller promises to deliver him the grain in a few days time. Where is he to put his bleating "money", which also needs to be fed, watched and tended? What if they should get the plague? And what if he should need to buy a knife, for example, which costs a third of a sheep? After all, he wouldn't cut a sheep into three pieces.

Metal Money

It is easy to imagine the properties of a commodity capable of performing the functions of money well. First, it must be homogeneous and of equal quality. One unit of money must not differ from another, like a well-fed sheep from a scraggy one. Secondly, it must be durable and lasting. For example, fish goes bad fairly quickly no matter how well it is salted or smoked. Thirdly, it must be economically divisible. If, say, sable was money, one-tenth of a pelt would not be worth one-tenth the price of a complete pelt. Fourthly, it must be compact and of high value per unit of weight or counting. As ethnographers recount, on certain islands in the Pacific stone money was in circulation and it was very difficult to drag from place to place; this was very bad money, of course. Fifthly, and finally, the money commodity should preferably have a relatively constant value, not too dependent on the caprices of the weather and other random phenomena.

The total or partial absence of these qualities in all the above-mentioned forms of "money" is the second great difficulty of exchange. It was solved by the emergence of *metals* in the role of money. In Ancient Sparta iron money was in circulation. Copper retained the functions of money until the comparatively recent past. We have information about lead and tin money. But as money these metals were far inferior to the precious metals, gold and silver, which possess yet another quality of considerable importance: they are aesthetically attractive.

No commodity can become money if it does not possess a use value. In the days before people learned how to work iron, gold was used together with copper to fashion certain simple small and light artefacts. As primitive-communal system disintegrated gold and silver were used increasingly for the production of ornaments and luxury articles. Articles made of these metals became the distinguishing signs of tribal chiefs who were turning into patriarchal slave-owners.

At the same time one cannot fail to see the contrast between the monstrous power that gold is acquiring and the limited nature of its use value. Gold satisfies not man's basic needs—for food, clothing, shelter and tools, but his secondary needs, the development of which in society is engendered by such base human qualities as vanity, miserliness and love of power. This contrast can be felt in the Greek legend about King Midas whom the gods punished by giving him a terrible gift: everything he touched turned into gold. He became immensely rich, but had to go without food and the use of other simple objects of consumption.

Gold became money in a comparatively "rich" society which had come a long way from the primitive hunting and farming communities. In the developed slave-owning civilisations of Egypt, Mesopotamia, Asia Minor and Greece (3000-1000 B. C.) gold was already a symbol of wealth and power. In some cases it was paraded, in others hidden away in caches and hoards. Both reflect the increase in the economic and social functions of gold.

Meanwhile these functions grew more complex and changed. Together with the original functions of a *measure of value and means of circulation,* the function of *hoarding* acquired great importance, the accumulation and retention of value as congealed human labour in the most compact and socially significant form. The function of *means of payment* also developed and money became a means of granting loans and settling debts. Finally, with the development of contact between peoples the function of *world money* became important, that is, of a means of international settlements, the transfer of value from one country to another, a universal materialisation of wealth.

Before the precious metals became the predominant form of money, the main way of calculating money was *counting*: heads of cattle, the pelts of some fur-bearing animals, cocoa beans, etc., were counted one by one. Obviously this was rather inconvenient; the units of this money inevitably differed from one another in size and quality. Precious metals made it possible to go over to the *weight* type of money. This was facilitated by the natural physical properties of gold—its softness, malleability and ductility. Gold dust is also easy to weigh. Jack London's heroes, the gold prospectors of the late 19th century, paid in saloons with ounces of gold dust and to do so every self-respecting prospector had to carry a pair of scales with him.

Weight money of gold and silver was great progress, but it did not solve all the problems. Imagine that our potter receives a certain weight of gold in pieces, bars, wire, dust, etc. He or the buyer must have scales of great accuracy in order to weigh out the necessary quantity of gold. If they do not trust each other they will have to seek the assistance of an intermediary, a weigher. Moreover, the potter would have to be an expert in precious metals to check the purity (fineness) of the gold he is given. In its natural state gold is usually found with admixtures of silver, copper and other less valuable metals. To tell the fineness of the metal with your eye or your tooth is very difficult, and the potter risked being deceived. Here is a story from a traveller about how silver functioned as money in Burma in the 1860s: "When you go to the market in Burma ... you must equip yourself with a piece of silver, a hammer, a chisel and the necessary weights. 'What do your cooking pots cost?' 'Show me your money,' replies the vendor and decides from the appearance of it on the price in a certain weight. Then the vendor gives you a small anvil and you hammer off silver until you think you've found the right weight. You must weigh it with your own scales, because those of the vendor are not to be trusted, and you add or take away until the weight is right. Naturally a lot is lost through the splinters that fall off, and it is always preferable not to buy the exact quantity that you weigh, but the equivalent of the piece of silver that you have just cut off. In large purchases which are paid for in the finest silver, the process is even more complicated, in that you must first call an assayer to test exactly how pure the silver is and to be paid for so doing."[2] There are even more such difficulties with gold, because each unit of it costs far more than that of silver.

The Coin

The third big difficulty of exchange was that originally in the economy there were no precious metals in the form of ready-made, weight units authorised and guaranteed by an authoritative body.

It was solved by *minting coins.* For a coin is, in fact, a guaranteed amount of metal. If a person held in his hand a five-rouble coin of tsarist times, he could be sure that it was 87.12 *dolia* (3.871 grammes) of pure gold. If it was an English sovereign, it contained 7.322 grammes of pure gold.

The emergence and development of the coin took many centuries. At first the practice grew up of stamping ingots, which was a certain guarantee of weight and standard. We know that such silver ingots were used in Ancient Babylon in the 3000-1000 B. C.

Who can certify the weight and fineness of a metal ingot? Obviously only a guarantee by authoritative persons or institutions will be trusted. Merchant corporations, municipal authorities and churches acted as these guarantors. But already at the early stages the state took the matter into its own hands. The minting of coins became the privilege of the state which gained considerable profit from it.

Stamped ingots of metal can be regarded as the embryonic form of the coin. It was much later, however, before the coin appeared in forms similar to those of today. The minting of coins—metallic discs or plates of different shapes with various representations and inscriptions—required a relatively high technology and art. The question of where and when the first coins appeared has not yet been fully answered. But we have no reason to disbelieve Herodotus, who tells us the following interesting details about the Lydians, a people who inhabited the western part of the Asia Minor peninsula: "The Lydians have very nearly the same customs as the Greeks, with the exception that these last do not bring up their girls in the same way. So far as we have any knowledge, they were the first nation to introduce the use of gold and silver coin, and the first who sold goods by retail. They claim also the invention of all the games which are common to them with the Greeks."[3]

As we can see, Herodotus is careful to qualify his statement on the minting of coins in Lydia ("so far as we have any knowledge"). He clearly does not believe that the Lydians invented the Olympic Games, but he accepts the information about how they bring up their girls.

The beginning of minting coins in Lydia is usually dated to the 7th century B. C. In the following century it evidently spread to the whole of Greek civilisation in the Mediterranean, and also to Persia which had emerged as a power by that time.

Gold was so firmly established in the minds of the ancients as the embodiment of wealth that the Lydian King Croesus (6th century B.C.) was considered the richest ruler precisely because he possessed a lot of gold. Preparing to fight the Persians, Croesus sacrificed in the temple of Apollo in Delphi unprecedented riches, which included 117 bars of precious metal, of which four were gold, weighing 2.5 talents* each, and the rest were a fusion of gold and silver, weighing 2 talents each, and also the statue of a lion made of pure gold weighing 10 talents.[4] This obviously amounted to several tons of gold alone. But this rich sacrifice did not save Croesus and his kingdom. Perhaps it even accelerated the catastrophe: the Persians undoubtedly knew about King Croesus' sacrifice of gold, and this must have whetted their appetites even more. After all, Croesus had not sacrificed all his gold to the god! Herodotus tells us that after taking the capital of Lydia the Persians seized a large amount of gold.

By the time of the formation of the mature slave-owning civilisation in Ancient Greece and Rome gold and silver were the predominant form of money, copper played a secondary role, and minting was on a fairly high technical level. This can be seen from the large number of antique coins found in museums and numismatic collections.

The fall of Rome and the conquest of Western Europe by peoples at a lower stage of development brought about a reduction in trade and money circulation. Early West European feudalism showed an interesting return to a natural economy. The mining and minting of gold almost ceased for several centuries. But the economic growth in many parts of Western Europe in the 14th to 16th centuries demanded an increase in the amount of gold and silver. The development of capitalism in the following centuries saw the penetration of money relations into all the pores of society. Money became the main means of social relations, the main aim of production and economic activity. Whereas the feudal lord and the serf are bound by relations of mutual dependence, there is only one link between the capitalist and the wage worker—the money which the former pays for the labour power of the latter. Capital acts primarily as a certain sum of money capable of producing new money.

Gold and silver coins were the predominant form of money in early capitalism. The growth of industry and trade demanded

---

* The Attic talent was equal to 26.2 kilograms. This was at the same time a measure of weight and the largest monetary unit in Greece and the neighbouring countries.

stable money circulation, stable currency. The development of minting coins was also promoted by the gradual disappearance of small feudal units and the formation of strong centralised states in England, France, Spain and some other countries.

Two Metals

All through recorded history gold and silver in the role of money appear as partners and rivals. Gold has always been more expensive than silver: this is explained by the fact that because of the natural conditions of its occurrence the production of a certain amount of gold requires several times more labour expenditure than that of an identical amount of silver. This value ratio of the two metals is constantly changing. This is because the labour expenditure in producing each of the metals changes, and with it the production costs. The intensity of the demand for gold and silver is also an important factor.

For many centuries, right up to the Classical Age of Greece, the ratio between the value of gold and silver fluctuated between 6:1 and 8:1. In Greece in Herodotus' time it was about 13:1. In the Rome of the end of the Republic and the first emperors the value of gold was eleven to twelve times higher than that of silver. Subsequently gold continued to grow relatively more expensive, and by the 5th century A.D. in Western Europe the ratio of the values of metals had reached 18:1. Then, by the 15th century, it dropped to 10:1, but in the 18th-19th settled at the level of 14:1 or 15:1. At the end of the 19th century silver began to drop in value quickly, which helped to drive it out of the sphere of money circulation. Silver gradually ceased to be money and turned into an ordinary commodity. It can be said that by now this process has been completed.

Whereas the dollar price of gold was strictly fixed until comparatively recently, the price of silver in the 20th century has changed constantly and dramatically. Consequently the ratio between the two metals has also changed, but this has been of little importance for the monetary system. In 1975 on average (on the basis of the market dollar price of gold and silver) it was about 38:1.[5]

In 1979-1980 silver experienced some dramatic ups and downs. Its price rose under the influence of universal inflation and the search for even more stable values than that of gold. As a result the price ratio of the two metals changed in favour of silver to become approximately 20:1. But then the silver market suddenly collapsed and its price dropped again. Insofar as the market prices of the two precious metals are fluctuating greatly, the ratio between them will undoubtedly continue to fluctuate within broad margins.

For two thousand years coins were minted from both gold and silver. Acting in this way, states were forced to fix the value ratio between the two metals formally. Thus, a dual universal equivalent and a dual scale of prices were forced upon the market, commodity turnover and the economy. It was assumed that a certain amount of silver was simply one-tenth or one-fifteenth the value of an identical amount of gold. But this contradicts the very nature of the universal equivalent.

Whatever the ratio fixed by the state, sooner or later it ceases to be a real one and the market fixes another, real ratio.

Both theory and practice have shown that gold and silver money cannot circulate together for any length of time. One or the other is driven out of circulation. Which one? The "more expensive" money, for which the law has fixed a price lower than the market price. It is believed in the history of money that this phenomenon was first noted by the English statesman Thomas Gresham in the 16th century and is therefore sometimes referred to impressively as Gresham's Law.

The mechanism by which the "underpriced" metal is driven out of circulation is as follows. Let us assume that the ratio of gold to silver has been fixed by law at 15:1, but the ratio 20:1 is operating on the market. Gold, as we see, is underpriced by law, it costs more on the market. In such a state of affairs people will try to get the state to exchange their silver for gold, then hide it, melt it down, take it abroad, and make all their payments in silver. As a result gold will be hoarded. If the official ratio is 15:1 and the market ratio 10:1, the reverse will happen: silver will be hoarded.

However, these particular fluctuations take place against the background of a long-term tendency. The more valuable metal drives out the less valuable. First copper ceased to be used as money, then silver. In the latter half of the 19th century and the beginning of the 20th gold reigned supreme on the "throne" as metal money. In most of the developed capitalist countries the monetary system of *bimetallism* (the two metals of gold and silver) had been replaced by that of gold *monometallism* or the *gold standard*.

Throughout the centuries gold was the money *par excellence*. Even during the period of bimetallism people's ideas about money and the social phenomena characteristic of it were linked more with gold than with silver. Later the equation of money with gold became even closer.

Gold and Its Substitutes

The transition to the gold standard extended over a fairly long period and was not completed in the main capitalist countries until the end of the 19th century.

Gold took over all the functions of money. Prices were now expressed only in gold, insofar as the monetary unit of each country was an amount of gold fixed by law. Gold actually circulated in the form of coins, and the circulation of these coins was constantly growing in terms of absolute amount. The central banks, in which the national monetary wealth was amassed, became the custodians of gold reserves. But among private individuals from the wealthy strata the custom of keeping gold at home in the form of coins or simple articles became firmly established. Gold could pay off any debt and was the means for final payment. Finally, the system of the gold standard became a world-wide one, embracing all the most important countries and their colonies. Gold was not the only form of money in this system, but it was the hub of the whole monetary system, the main form of money to which all other forms were reduced.

Yet the more firmly gold became established in the role of metal money, the more inexorably the conditions for its driving out developed. Driving out by what? By another metal? No, there was no more suitable metal for the role. We are talking of the driving out of gold by *paper-credit money*.

It all began with the fact that gold was not indispensable for the function of a means of circulation in all its lustrous brilliance. If the potter sells pots only in order to buy wheat and oil straightaway, he is not interested in gold as such. It is important for him to obtain some kind of "guarantee" which the owners of wheat and oil would readily accept. If the latter, in their turn, are selling their goods only in order to buy clothing, footwear or, say, wine, the lustre and weight of a coin is not of any special significance for them either.

This important point revealed itself first in the fact that alongside coins of full weight worn coins that have lost part of their weight circulate with equal success. And there are always some artful dodgers who begin to use this economic law to their own advantage. They cut pieces off coins with the result that the weight of metal contained in them is reduced. Together with the profession of minting that of coin forging arises. The state is in a particularly advantageous position to profit from the issue of under-weight and inferior coins. The history of money circulation is one of the "spoiling" of coins, that is the issue of increasingly light-weight coins or coins with a higher content of cheaper metal which retain the former face value.

The gradual evolution of money from leather and sea-shells to gold is strangely combined with the reverse tendency—from gold to cheap "ersatz". But this is strange only at first glance. The substitutes are not money in the old, commodity sense of the word; their value is not determined by the expenditure of labour on their production. Now they are token money, symbolic money.

The driving out of commodity money by token money took many centuries. It was not a smooth, steady process, and it has not been completed even now, and cannot be completed as long as capitalism exists.

It would seem that the laws of money circulation are turned upside down by the introduction of token money. Commodity money could circulate because it had the value determined in the final analysis by the expenditure of labour on its production. Token money, on the contrary, acquires its "value" (purchasing power) because it circulates.

Not counting cut or inferior coins, the first symbolic money in history is thought to have been made from leather and cloth. But then why not make it out of paper, a material that is in a sense itself a symbol of the civilisation of the modern age? Paper money can be regarded as a type of credit money. For its circulation is based on confidence in it, that is, on the credit of the body issuing it. Gold does not require confidence, it speaks for itself. But paper money circulates because people have confidence in the institution or authority that issues it. Such an institution or authority is more often than not a state treasury or central bank. People who have paper money in their pocket or safe are thereby giving credit to the institution that puts this money in circulation. In other forms of money (current accounts in banks or savings banks) its credit nature is even more obvious.

Inferior coins made of copper, nickel, aluminium and even silver is credit money in the economic sense of the word: it is accepted not because of the value of the metal, but because the state has put its stamp on it. True, inflation and constant rise in the price of these metals have created in many capitalist countries a strange position with small change. The silver coin has disappeared from circulation, because the value of the metal in it began to exceed the coin's face value. Moreover, silver coins are often of numismatic value. The silver dollar has ceased to be credit money, but one might say that it has ceased to be money at all and has turned into a specific commodity. It may be worth ten or twenty paper dollars depending on the market price of silver metal, the rareness of the issue and a number of other factors. Something similar is happening to money made of non-precious non-ferrous metals. The Director of the State Mint in Munich, West Germany, told me in April 1979 that it cost the state 3 pfennigs to mint a 1-pfennig coin. The normal position, in which the state obtains a certain profit from minting coins, has turned into the reverse. In order to provide the economy with small change, the state is forced to incur these losses; they are probably covered by the profit which it still makes from minting coins of higher denominations. But the losses on pfennigs are growing, because the economy is devouring them in vast quantities. This can mean only one thing: someone is melting them down and making a profit from it.

Let us return, however, to the question of the history of the development of paper-credit money and its relationship with gold.

Paper-credit money first appears as a substitute for or representative of gold. If not too much of it is issued and if it is exchanged for metal at a fixed rate, this money can circulate successfully and be of considerable use to the economy. Up to a certain stage the driving out of gold by its substitutes is the progressive development of the gold standard.

According to the American economist Robert Triffin, the structure of the money supply in terms of value in the three most developed countries—the USA, England and France—in 1815 was as follows: gold coins—33 per cent, silver coins—34 per cent, ready paper money and small change (copper)—26 per cent, money in bank accounts—6 per cent. As we can see, this was still the period of bimetallism and underdevelopment of the credit system.

In a century, by 1913, important changes had taken place in the structure of the money supply. The share of gold dropped to 10 per cent, which, however, did not prevent it from being at the height of its power during this period, the hub of the whole monetary system. The share of silver dropped to 3 per cent, and it began to play a subordinate role. Ready money (paper and copper) accounted for 19 per cent, and the remainder—68 per cent—was made up of money in bank accounts. The credit system had developed tremendously.[6]

Very early (in the 17th century at least) governments began to issue paper money, which was not exchangeable for gold or silver. It was particularly easy to do this if the public could be persuaded that it was necessary for financing a war. In the 19th and 20th centuries the state used the credit system for the same purpose, forcing banks to accept its bonds in exchange for entries in bank accounts to the credit of the treasury. The result of these financial operations is well known: the depreciation of paper-credit money in relation to metal and the general rise in commodity prices, that is, *inflation*.

In modern conditions paper-credit money has lost all connection with gold. It has become the only functioning form of money. Its depreciation can be measured by the growth of the general level of commodity prices, which is expressed with the help of various indexes. At the same time this money usually depreciates in relation to gold also, which can be seen in the rise in the price of gold. But the economic and social importance of this process is now relatively secondary. The rise in the price of oil is, perhaps, a more important element of inflation than the increase in the price of gold.

If gold is the yellow devil, then paper money is regarded by many as diabolical invention, because it is an instrument for robbing poor and ignorant people. Goethe portrays paper money as such in the Part II of *Faust.*

What was in Goethe's time an exception, has now become universal practice. It is now several decades since gold coins went out of circulation in capitalist countries. Quite recently the last vestiges of the convertibility of paper-credit money into gold vanished. Inflation has become a constant element of capitalist economy. It so occupies the minds of economists and politicians that they tried to find out how the actual term arose. It has been established more or less definitely that it was first used in English in 1864 by an American called Alexander Del Mar, the author of some interesting works on the history of gold and money.[7] This was in the period of the Civil War in the United States, when the Washington government issued a large quantity of inconvertible paper money, the so-called "green-backs". In Russian pre-revolutionary literature the word "inflation" was hardly used and it first appeared in 1923-1924, particularly in connection with the colossal depreciation of the mark in Germany.

Can paper-credit money not connected with gold serve the economy well and retain stable purchasing power? It can, but only in effectively planned economy, where the issue of money and money incomes are flexibly coordinated with production and the realisation of goods and services. Only socialism creates such conditions. But in socialist society too this possibility is realised not of itself, but by overcoming considerable difficulties.

The improvement of the monetary system under socialism should proceed through the development of paper-credit money not connected with gold. The stability of its purchasing power is guaranteed by the steady growth of social production and commodity turnover and improved planning. The expansion and improvement of bank transfers are of great importance. The proportion of bank money and cashless payments is increasing and will continue to do so at the expense of cash paper money. The coming decades will be a period of increased automation of bank settlements and trade, accounting and document processing.

III. MINING THE YELLOW METAL

The author of witty aphorisms and fables, Felix Krivin, wrote the following on our subject:

"Oxygen is essential for life, but it is also no easy matter to live without Gold. What really happens? When it is easy to breathe and all is well with Oxygen, you haven't enough Gold. But when you are deluged with Gold, it becomes hard to breathe, and that means there isn't enough Oxygen.

"For according to the laws of chemistry—the most ancient laws of Earth—Gold and Oxygen do not combine."[1]

The moral of this reasoning is worth reflecting upon. Let us recall the last words of Pushkin's Covetous Knight, the possessor of six chests of gold, who became the victim of his own wealth:

'Tis faint I feel... My legs give under me...
I cannot breathe... Air! Give me air!...
My keys! Where are they?... My keys!...[2]

This moral applies not only to the owners of such exotic treasures, so remote from everyday life, but also to accumulators of far more prosaic forms of property... Here, as in many other cases, gold is a symbol of wealth, a symbol of property.

As for the chemical phenomenon used here as a metaphor, it is beyond all doubt. As we already know, the fact that gold does not oxidise has played an important role in its economic functions and still does.

Before discussing these functions in more detail, we must say a word, albeit briefly, about the form in which gold exists naturally, where it is obtained, and how it acquires its socio-economic functions.

Chemistry, Physics and Geology

In the encyclopaedia we read: *Gold* (Latin *aurum*), Au, a chemical element of the first group of Mendeleyev's periodic system; atomic number 79, atomic weight 196.9665; a heavy yellow metal. Consists of one stable isotope.

Together with copper and silver it forms in the first group a special second subgroup of elements which, by virtue of common features in their atomic structure, possess many similar chemical properties. This triad is also linked by geological affinity: gold exists naturally together with silver and in many cases copper.

The chemical and geological affinity of gold with silver and copper undoubtedly lies at the basis of their economic affinity: all three metals have performed the functions of money. Gold and silver are often used together in jewelry-making.

The chemical inertness of gold has exceptions. Insoluble in most acids, it is helpless before "tsar's vodka", a mixture of hydrochloric and nitric acids. The solubility of gold in sodium cyanide NaCN and potassium cyanide KCN is the basis of the cyanidisation process, now the chief method of extracting precious metal from ore. The introduction of cyanidisation at the end of the 19th century revolutionised the southern African gold-mining industry, by making it possible to extract the metal from ores with a comparatively small but unusually stable metal content.

The physical properties of gold are also important for its economic functions. Gold is a very soft, plastic metal. It possesses almost unlimited malleability and can be hammered into an extremely thin, almost transparent pellicle and wire thinner than a human hair (1 gramme of this wire stretches two kilometres). Gold is a good conductor of heat and electricity and melts at a temperature of 1,064° Centigrade. Its density (specific gravity) is 19.3 grammes per cubic centimetre. Only metals of the platinum group are heavier than gold.

Standard bars of monetary gold alone consist of almost chemically pure metal (about 99.5-99.9 per cent). In technology, jewelry-making and minting coins gold is always used in alloys, more often than not with copper and silver. These admixtures give it durability and other necessary properties, and also make the articles cheaper. The gold content in the alloy is called fineness.

In the USSR and many other countries the fineness is expressed by the number of parts of gold per 1,000 parts of the alloy. The fineness of the above-mentioned bars is from 995 to 999. Until 1927 the old Russian measure of fineness existed in the USSR, designated by the number of zolotniks of pure gold in a pound, which was divided into 96 zolotniks. Finally, in certain Western countries, particularly in Britain and the United States, the carat fineness is still used, which is expressed by the number of twenty-fourths (carats). For example, for some articles of jewelry 958 fine gold is used, which corresponds to 92 in the zolotnik system and 23 carats. As a rule gold of a lower fineness is used for articles.

The English sovereign was minted, and still is for export, from 22 carat gold, which corresponds to 916.67 fineness. With a content of pure gold of 7.322 grammes a coin weighs 7.988 grammes. Naturally it now costs not one paper pound sterling, but a great deal more.

The American gold coin with a face value of twenty dollars (the so-called "double eagle") was minted from gold of 900 fineness and weighed 33.44 grammes, whereas the gold in it weighed 30.093 grammes. The heaviest coin in modern times was the Mexican centenario (50 pesos), which weighed 41.66 grammes and contained 37.494 grammes of gold. This giant was 37 millimetres in diameter and 2.8 millimetres thick.[3]

Gold is everywhere: in all types of soil, all ores and in sea water. But in the vast majority of cases its concentration is insignificant. According to recent estimates one ton of matter of the earth's crust contains on average about 4 milligrams of the yellow metal.[4] This is thousands of times less than the content of copper, zinc and lead. It is interesting that meteorites are usually far richer in gold than the earth's crust. Since it is assumed that the composition of meteorites is similar to that of the central parts of our planet, some scientists have advanced the hypothesis that together with other heavy metals there is a large amount of gold in the earth's core.

Metals are usually extracted from minerals that are chemical compounds containing the metal in question. Although gold forms about 20 different minerals, only *virgin gold* is of commercial importance. This is embedded crystal grains of different sizes, ranging from specks invisible to the naked eye to nuggets weighing tens of kilograms. As one geologist puts it, "Relatively large gold grains—'visible gold'—are rarely found, and a geologist who has worked for many years in a rich gold ore deposit may never see specimens containing grains."[5] In Russia a decree was issued as early as 1825 to the effect that all nuggets found should be handed over to the state museum, thereby laying the foundation for the now famous collection of the Diamond Fund of the USSR.

There are only a few dozen large nuggets in the whole world and they each have their own name. The largest one in Russia, the Big Triangle, weighing about 36 kilograms, was found in the Urals in 1842 and can today be seen in the Diamond Fund of the USSR. Many large nuggets found in other countries have unfortunately shared the fate of works of art made of gold: they have been melted down for the metal. This is why we possess little information about the largest nuggets. According to some statistics, the biggest is the Welcome Stranger weighing 70.9 kilograms with a content of 69.6 kilograms of pure gold, found in 1869 in the rut of a country road in Australia. The German scientist Heinrich Quiring, author of one of the most definitive works on the history of gold, writes that a nugget weighing 193 kilograms was found in Brazil around the middle of the 19th century.[6] More reliable is the information about a piece of a gold lode called the Holtermann Nugget found in Australia, which weighed 285 kilograms together with the rock and contained 93 kilograms of gold.[7]

But the vast majority of gold is extracted, of course, not in the form of nuggets. Two main types of deposits are of commercial importance—primary (ore) and secondary (placer or alluvial) deposits. The first were formed in the course of ancient geological processes, when the formation of the earth's crust took place, and are usually found in mountainous regions. The second are the result of the action of wind, water and other natural forces on gold ores and are generally found in the courses of existing or former rivers. Ore deposits are frequently found at great depths and need to be extracted by mining. Placers are usually closer to the surface or actually on the surface, but sometimes are covered by fairly strong layers of empty rock.

Recently scientists agreed to distinguish a third type of deposit, where the gold has been formed as a result of the repeated action of geological processes on previously destroyed rock. Examples are the famous South African conglomerates of Witwatersrand. But in the conditions of their bedding and extraction they are very similar to ore deposits and are often classified in the same group.

Naturally enough, people first began to extract gold from placers: they are more accessible and do not require complicated work and equipment. The Californian, Australian and Alaskan gold rushes of the 19th century and the many rich discoveries in Russia were associated mainly with placers. According to present estimates, 90 per cent of the world production of gold in 1848-1875 came from placers. But in 1929 this percentage dropped to 8 in the capitalist world, and in 1971 to 2 per cent.[8] Almost all gold is extracted from the conglomerates of the Republic of South Africa and from primary deposits in various parts of the world. In recent years the percentage of gold extracted as a by-product from silver, copper, uranium and other ores has increased considerably.

In 1923 in the USSR on the River Aldan in Yakutia a new deposit of precious metal was discovered, placers which are still being exploited today. In the Soviet Union the extraction of gold from placers with the use of modern highly productive technology is still of considerable importance.

The discovery of a new large gold deposit is always an exciting, sometimes a dramatic and even a tragic event. Descriptions of real and imaginary discoveries are frequently found in literature. Here is one episode written by a great master (Jack London in the story *All Gold Canyon*):

"It was only half rock he held in his hand. The other half was virgin gold. He dropped it into his pan and examined another piece. Little yellow was to be seen, but with his strong fingers he crumbled the rotten quartz away till both hands were filled with glowing yellow. He rubbed the dirt away from fragment after fragment, tossing them into the gold pan. It was a treasure hole. So much had the quartz rotted away that there was less of it than there was of gold. Now and again he found a piece to which no rock clung—a piece that was all gold. A chunk, where the pick had laid open the heart of the gold, glittered like a handful of yellow jewels, and he cocked his head at it and slowly turned it around and over to observe the rich play of the light upon it."[9] The story's plot develops as follows: the prospector is attacked by a robber, but with the help of cunning the lucky man overcomes and kills his opponent and, loaded with gold worth 40,000 dollars, sets off home chanting a psalm.

Remarkable discoveries of gold still take place in real life in our day. But the fate of gold production does not depend on them, of course. Geological prospecting for gold is now carried on on a strictly scientific basis and with the help of complex and expensive equipment. Its success depends on the coordinated work of many specialists. In the course of prospecting the boundaries of the deposit are ascertained, the average gold content in the ore is estimated, the conditions of the gold deposit are taken into account, and economic estimates are made as to the profitability of extraction.

How It Is Extracted

Timothy Green, who has conscientiously travelled round the world in order to experience gold with his eyes, ears and hands, perhaps even nose and tongue, in all its forms and at all stages of its movement, describes a modern South African gold mine as follows:

"Going down a gold mine is rather like a trial run for Hades. You even leave all your clothes, including underwear, behind on the surface and, shrouded in white overalls, enter a steel cage which plummets through a mile of rock in two minutes. There below is a noisy, hot, wet world lit by the dancing fireflies of the lamps on miners' helmets. A ten-minute walk along a gallery cut through rock whose natural temperature is over 100° Fahrenheit, and any visitor is soaked by a combination of sweat and humidity. Then, above the constant hum of the air conditioning and the rumble of trucks along steel rails, comes the sound of compressed air drills biting into solid rock. On one side of the tunnel a narrow opening begins plunging down at an angle of nearly 25 degrees towards the bowels of the earth. It is barely 40 inches high and is delicately held open by props of blue gum. It is called, in mining parlance, a stope. Within the stope the rock seems to press in from all sides; tiny flakes fall from the roof into the pools of warm water in which everyone is kneeling or lying. Almost hidden in a fine spray of water to subdue dust, the long needle nose of a drill chatters into a hole in the rock marked with a blob of red paint. All along the side of the stope a continuous line of red paint highlights a 4-inch vein of rock that, even to the uneducated eye, looks markedly different from the rock above and below. It is a tightly packed bunch of white pebbles and between them, here and there, a minute speck of gold gleams in the beam of the miner's lamp. This vein or reef is the meat in the sandwich. This mine, Free State Geduld in the Orange Free State, is one of the very few in which the gold between the pebbles can actually be seen by the naked eye, for it is blessed with one of the richest reefs ever discovered in South Africa."[10]

With an average gold content of 26.4 grammes per ton of ore, the Free State Geduld Mine processed about two million tons of ore in 1971, from which 53 tons of the metal was extracted. This content was double the average for South African mines. The highest number (at the West Driefontein mine) was 31.4 grammes per ton. This mine yielded 89 tons of gold. These two huge mines had the lowest production costs. They were respectively 14.1 and 15.3 dollars per ounce of gold at the then current market price of about 40 dollars per ounce.[11] Large-scale production in favourable natural conditions yields good results and high profits.

Mines like the ones in South Africa are hardly found in any other gold-producing country. The modern mine is a large, complex and very expensive enterprise. The construction of such a mine takes a long time (usually five to six years) and enormous capital investment. At the beginning of the seventies the building of a mine cost at least 100 million dollars. We can assume that the cost has doubled or tripled since then. At great depths, now sometimes as much as 3,000-3,500 metres, the cooling, disposal and supply of water, ventilation, removal of obstructions, energy supply, and horizontal and vertical transport pose complex technological problems.

The companies incur only the most essential expenditure needed for production and do not seek to lighten the miners' work. The Republic of South Africa is far behind the USA and Canada in mechanisation of extraction and in labour productivity. In this connection Green writes: "Actually the gold-mining industry has been very slow over the years to develop mechanisation underground, because it has so frequently been cheaper just to send in another ten Africans, instead of inventing a machine."[12] The miners' work in the South African mines is hard and dangerous. Each of the largest mines employs up to 20,000, of whom usually about 90 per cent are Africans. In underground work the percentage of Africans is even higher, because there the whites perform only managerial and technically complex work. In particular, Africans are not usually granted access to explosives and only whites are employed in blasting operations.

The extraction of ore underground and its delivery to the surface is a most important, but only the first stage in the technological chain of gold production. The processing of the ore and the melting of the metal are done at complete-cycle factories. In the structure of capital expenditure on the construction of a typical mine in South Africa this expenditure accounts for about 15 per cent. About 6 per cent of the total labour force in the industry is employed at factories.

The ore passes through several stages of crushing, grinding and sorting. Gold is extracted from the pulped ore by the cyanide process or other methods. Melting at mining factories in South Africa yields rough bars of gold weighing about 1.000 ounces (about 31 kilograms) and containing from 85 to 90 per cent pure gold. The final refining is done by electrolysis at a central plant (the Rand Refinery) which belongs to the Chamber of Mines of South Africa. This is an influential association of entrepreneurs which performs many important economic and political functions, in particular, recruiting workers for the mines and representing the industry in its relations with the government.

From the refinery the gold emerges in the form of standard ingots (bars). These bars usually weigh 402 ounces (about 12.5 kilograms) and are 996 fine. International standards traditionally permit bars that weigh from 10.9 to 13.8 kilos and are not less than 995 fine. These bars provide the primary raw material for all the subsequent uses of gold.[13]

The main advantage of working placers is that it does not require deep mines, and is more often than not done by the open-cast method. There is usually no expenditure on crushing ore either. Engineering has done a great deal to mechanise and increase the labour productivity in the extraction of gold from placer deposits. Today use is made in this sphere of dredgers, hydro-monitors (water guns to wash auriferous sand), slurry pipes, excavators, bulldozers and other machinery. The further processing of auriferous pulp and the extraction of gold are practically the same as those with primary gold.

We know that right up to the 1970s the price of gold in the world market was fixed at an artificially low level. In most countries the extracting of placer gold could not withstand the competition of the South African mines and was gradually dying. It is possible given the new level of prices that the mining of certain placer deposits will be resumed and expanded, but specialists do not expect any significant changes.

In recent years a more serious rival for the South African mines has emerged, namely, the obtaining of gold as a by-product mainly in the production of non-ferrous metals. In Canada in recent years about one-third of the total gold output comes from incidental output, and in the USA approximately 40 per cent. In the 1970s the young island state of Papua-New Guinea in the Pacific Ocean gained an important place among the gold producers. On the island of Bougainville, which forms part of it, there are some very rich copper deposits which large foreign companies have begun working. They are also gold producers. In 1978 Papua-New Guinea ranked fourth among the gold-producing non-socialist countries, with an output of more than 23 tons. It can be assumed that the total incidental production of gold in all these countries has now reached about 100 tons. This is quite a large amount, but nevertheless it is seven or eight times less than gold production by the South African industry, which retains a virtual monopoly in the capitalist world.

How Much, When and Where

Statistics know everything. But—and this must always be remembered—with varying degrees of reliability. The question of how much gold has been and is being produced in the world began to interest people long ago, but it was not until the 19th century that some more or less realistic estimates of past production were made and only at the end of the century did statistics become satisfactory. Without claiming any great accuracy, we can assume that in the whole history of gold production, which covers more than 6,000 years, mankind has dug up from 90,000 to 100,000 tons of gold out of the earth.

Is that a lot or a little? A lot, if one takes into account that its extraction is highly labour-consuming. Even today in the gold-rich Republic of South Africa 400,000 miners equipped with modern technology "hoist" 700-800 tons of pure metal annually, which means an average of about 2 kilograms per worker. And what hard work each tiny grain of gold gave the prospector equipped with nothing but a spade and pan!

But, in spite of the technological progress, compared with other metals long known to mankind, this is not very much, and in a certain sense it is very little. All the gold produced by mankind could be put in a cube with a rib of about 17 metres or, say, in an average-sized cinema. The amount of gold produced annually would fill only a small living room.

Incidentally, on the subject of total and annual production: nobody is particularly interested in how much oil has been produced or how much steel smelted since the beginning of time. This is not very important in relation to copper or even silver. But gold is a special case. Oil disappears in the process of consumption. Iron and steel go only in part for resmelting as scrap. The secondary use of copper and particularly silver is more important. But only gold is "everlasting": by virtue of its natural and social properties once it has been produced it does not disappear, does not vanish into the ground, water or air. It is possible that your wedding ring is made of gold produced three thousand years ago in Egypt or three hundred years ago in Brazil. This gold may have taken the form of a bar, a coin, a brooch, a cigar case and goodness knows what else in its time.

It is, of course, an exaggeration to say that gold is "everlasting". A certain percentage is "consumed" never to return. All melting down and processing of gold involves losses. When gold circulated as coins they were worn down from being handled by thousands of greedy or indifferent hands. Who knows how much has perished on the sea bed and in troves which no one will ever find? Finally, modern forms of the technical use of gold are partially destroying it in the sense that its secondary use is impossible or uneconomical (pellicles, solutions, medicaments, etc.). In the 1940s an American writer estimated this irrevocably lost gold at 7,000-8,000 million dollars, which corresponded at that time to about 6,000-7,000 tons (about 10 per cent of total output).[14]

Let us accept 10 per cent. Even so this means that nearly all the gold extracted is *economically active,* that it is in a form suitable for further use. Annual production adds only a very small percentage to the gold reserves accumulated by mankind: in recent years about one per cent. No other commodity comes anywhere near gold in this respect.

Naturally the most doubtful estimates of gold output are for the Ancient World and the Middle Ages. Quiring has made extremely careful calculations, using the evidence of ancient writers, documents, geological data and—perhaps most of all—his own intuition. He believes that up to the discovery of America about 12,700 tons of gold had been extracted in the world.

In the Ancient World (up to the fall of Rome in the 5th century) the two main gold-producing areas were Egypt (together with what is now Sudan) and the Pyrenean peninsula. Many monuments of material culture and writing have come down to us from the Egypt of the pharaohs which testify to the role of gold in its economy, the progress in the technique of extracting and melting gold, and the terrible conditions of slave labour in the mines. The artistic treasures from the tomb of the Pharaoh Tutankhamen (14th century B.C.) include some exquisite gold articles and are world famous.

Gold from Egypt made its way to the neighbouring countries of the Middle East. The diplomatic correspondence of the Egyptian rulers with the Kings of Babylon, Assyria and other neighbouring states (15th-14th century B. C.) found in the El Amarn archives contains numerous references to gold. The kings are constantly asking the pharaoh to share this wealth with them, flatteringly exaggerating its dimensions and offering various gifts in return. Such was diplomacy, foreign trade and international payments 3,500 years ago.

Quiring assumes that one of the inscriptions of the Tutankhamen period contains the name of the man who can be considered as the first geologist and student of minerals. A certain Renny reports that he was sent by the government to look for gold ores. It is highly likely that mining was taught in the ancient "university" at the temple of the god Ptah in On (Heliopolis).[15]

Gold mining on a large scale was also carried on in the area which is now Spain and which became the main source of gold during the age of the Roman Empire. The Romans built complex engineering installations for the crushing and washing of gold-bearing ores. A similar scale of work was not reached again until the 19th century. According to Pliny the Elder, the scholar and writer who was a highly-placed Roman official in Spain in the first century A.D., the mines in this area produced (translated into modern units of weight) about 6,550 kilograms of gold annually.[16] This is quite a lot even by present-day standards: such gold-producing countries as Mexico or Zaire have been mining about the same amount in recent years.

The Middle Ages in Europe was a period of the decline of gold production. Many techniques that had spread during the age of Rome were forgotten. The production of primary gold ceased altogether, and people only "washed gold" here and there in the beds of rivers and streams in a primitive way. Early Christianity praised poverty as a virtue and preached against gold. No gold coins were minted anywhere in Western Europe from about the 9th to the 13th century.

In the late Middle Ages the main gold-producing area was Tropical Africa. Enterprising travellers and traders, particularly Portuguese, set off in search of gold exploring further and further south along the west coast of Africa. The present independent state of Ghana was called the Gold Coast in colonial times, the name given to it by Europeans in the 15th century. The enterprising Robinson Crusoe, the hero of Daniel Defoe's famous story, gives natives somewhere on the coast of the Gulf of Guinea some trinkets in return for 5lb 9oz of gold dust (about 2.5 kilograms).[17]

The skilled and the scholarly sought to solve the problem with the help of alchemy, to find a way of obtaining gold from less valuable metals. Polymetallic ores usually contain a certain amount of gold invisible to the naked eye. When the ore was melted the gold separated from it and looked as though it was being made from silver or copper. This was the origin of both well-meaning mistakes and brazen cheating. As we know, the science of chemistry subsequently developed from alchemy. Alchemy greatly enlivens the history of the Middle Ages and the literature also, but, insofar as we are dealing here with statistics, we cannot credit it with a single gramme of produced gold.

There were many trends in alchemy, but certain general principles operated which had been introduced in the early Middle

Ages by Arabian alchemists. They believed that all metals arose as a result of the combination in varying proportions of sulphur and mercury. In this case the task of obtaining gold artificially was simply to find the right proportions and ways of combining these two initial materials. The alchemists believed that sulphur was the father of gold and mercury the mother.

It is interesting that, as modern nuclear physics shows, the theoretical possibility of obtaining atoms of gold from these two elements really does exist. To do so it is necessary to bombard mercury atoms with the accelerated nuclei of sulphur atoms. It is estimated, however, that this method of producing gold would be billions of times more expensive than the usual one. Other ways of obtaining gold by means of transforming elements are also theoretically possible, but unlike ever to be practicable.

These facts of modern science are in no way connected with the alchemists' "discoveries", of course, and are nothing more than coincidence. Mercury was probably regarded as related to gold because of its high specific weight, and sulphur because of its colour.

The belief in alchemy in the Middle Ages was so universal that Henry IV of England issued a decree forbidding anyone but the king to engage in transforming non-precious metals into gold. Yet already at a very early stage in the development of science there were people who said it was impossible to transform metals and that the alchemists' claims were nonsense. Among them was the great Oriental thinker Avicenna.[18]

The discovery of America opened up a new chapter in the history of precious metals. First and foremost, of silver. According to the estimates of Adolph Soetbeer, who is an acknowledged authority in the sphere of the statistics of precious metals output, the annual world output of silver exceeded in value that of gold right up to the 1830s, with a very large percentage coming from the New World.[19] This fact was of great importance for the fate of the monetary system: it prolonged the "monetary life" of silver and the prevalence of the dual system (bimetallism) up to the middle and the end of the 19th century.

The relative lagging behind of gold production in America is explained also by the fact that the Spaniards and Portuguese did not succeed in discovering any significant deposits of primary gold, and the working of placers could not provide stable and prolonged gold output. Nevertheless, right up to the discovery of gold in California and Australia in the middle of the 19th century South and Central America remained the world's main gold-producing region. According to Quiring's statistics, based on Soetbeer's estimates, in the 16th century America produced more than one-third of world output, in the 17th century more than half, and in the 18th century two-thirds.[20] In the 16th and 17th centuries gold was extracted mostly in Colombia and Bolivia, and in the 18th century in Brazil which held first place in the world during this period. Portugal, which then owned Brazil, was the first country to introduce a gold monetary system officially and to give up silver.

The end of the 18th and first half of the 19th century were "lean years" for gold output. The average annual gold output reached 24.6 tons in the period 1741-1760, then dropped steadily until it was only 11.4 tons in 1811-1820, and after that began to rise slowly.[21] It must be borne in mind that during this period the industrial revolution was taking place in Western Europe and North America. The production of commodities for sale grew rapidly, and therefore the need for money increased. Gold could not keep up with these events, and in the first half of the 19th century its future as the basis of monetary systems was by no means guaranteed. In the short period between the exhaustion of the South American placers and the Californian discovery Russia held first place in the gold-producing league. In 1831-1840 it produced more than one-third of world output and retained this leadership until the end of the 1840s.

Archaeological and historical data testifies that gold was mined in the Urals and the Altai in ancient times. The very word "Altai" comes from Turkic-Mongolian *altan*—meaning "gold". However, this mining was abandoned and forgotten, and the modern history of Russian gold begins in the middle of the 18th century, when it was rediscovered in the Urals. Later gold (predominantly in placers) was discovered also in Western and Eastern Siberia and the Far East.

In this "gloomy" period for gold the attention of the West was attracted to Russian gold mining by the great German scholar Alexander Humboldt, who took an interest in gold problems all his life. In 1838 he published a special work on the trends of world gold production, in which he included data about Russia that he had received directly from the Russian Minister of Finance, E. F. Kankrin.[22] It is possible that this information may have cheered up the bankers and economists in Western Europe somewhat.

A new—and extremely romantic—age in the history of gold began in January 1848 when, as Green writes, "a carpenter named James Marshall found what he thought were specks of gold in the tailrace of John Sutter's mill near the junction of the American and Sacramento rivers... At first Marshall and Sutter tried to keep news of the discovery quiet, but rumours of gold were not easy to quench and soon the word had spread to San Francisco, then a struggling port of about 2,000 people. By spring half California had deserted its farms and homesteads and rushed to the gold fields... By the autumn of 1848 the first rumours of the discoveries were flying around New York. Each day brought fresh news and the excitement mounted. What happened during the next few months was quite unprecedented in history. Thousands of men suddenly saw a spark of opportunity to earn a fortune in a matter of days... Even before President Polk finally confirmed the extent of the finds in a speech to Congress in December 1848, the scramble to get to the West Coast was on."[23]

John (Johann August) Sutter, mentioned above, was a remarkable person in his way. The first discoverer of Californian gold came from Switzerland and had settled in America only a short time before. He was an enterprising, energetic man, but inclined to romantic fantasies. His eventful life became the subject of a delightful historical miniature by the Austrian writer Stefan Zweig from the cycle *Sternstunden der Menschheit* (The Starry Hours of Mankind).[24] The discovery of gold on his land did not bring Sutter happiness. He died in poverty and obscurity. Sutter was not the first or the last of those whom gold finally destroyed, ruined and sent to their graves. We find such figures both in the true history of gold and in works of fiction.

When Zweig speaks of the Californian discovery as one of the "starry" hours of mankind, he has in mind the historical importance of this event. The discovery of gold in California, at that time separated from the centres of civilisation by vast distances not easily traversed, played an important role in the development of capitalism in the 19th century. The Californian gold which started to flow to the east coast of the USA and to Western Europe was like fresh blood for the capitalist economy. It stimulated the growth of industry, trusts, and large banks, the building of railways and the development of world trade. Gold played a particularly important role in the economic development of the United States. It promoted the settlement of new vast territories in the West and central regions, the economic drawing together of the individual states and territories, and the growth of a transport network.

The Californian gold rush with its human suffering and fates of simple people has gone down in the folklore of the American people and become the subject of countless literary works. But at the moment we are interested in figures. Thanks to the Californian finds the United States marched into first place in world gold production and retained it almost until the end of the century, giving way only occasionally to Australia whose own great gold rush began in 1851.

The discovery of gold in California and Australia and also the rise in output in Russia and some other parts of the world resulted in a dramatic change in the whole world situation with gold. In the second half of the 19th century 11,000 tons of gold were produced, eight times more than in the first, and double the amount for the whole period after the discovery of America.

Demand was growing even faster, however, for during this period the gold standard was introduced in all the main countries and gold became the basis of the monetary systems and the world money. Therefore by the 1870s, when the cream had been skimmed from placer deposits and no new large and easily accessible ore deposits had been found, pessimism began to spread among the specialists. In 1877 the Austrian Eduard Suess published a book entitled *Die Zukunft des Goldes* (The Future of Gold), which sounded most original at that time, but has become worn out by constant repetition and variations in the last century (the fate, prospects, chances, etc., of gold). Suess maintained, firstly, that future production of gold depended overwhelmingly on placer deposits; secondly, that mankind had already obtained more than half the gold accessible to it and the prospects for production were very unfavourable; and, thirdly, that there was nowhere near enough of the metal for the universal introduction of the gold standard.[25]

All these prophesies turned out to be mistaken, as did many later ones by the way. The history of gold abounds in false prophesies and wrong predictions. In 1935 the English financial expert Paul Einzig stated in a book with the same title *The Future of Gold* that, for example, "it is safe beyond doubt to take it for granted that the gold standard will never be universally abandoned". He also assumed that "it is beyond doubt that the demonetisation of gold would reduce its price to a fraction of its present figure".[26] Note that both statements were "*beyond doubt*". In fact the countries of the so-called gold bloc, which, led by France, retained the gold standard longer than all the others, had already abandoned it before the ink had time to dry on the pages of Einzig's book. The formal demonetisation of gold, which took place on an international scale in the 1970s, did not result in a drop in its market price and purchasing power.

In the 1870s and 1880s gold was on the threshold of the most brilliant stage of its career. The use of the dredger opened up new prospects for obtaining gold from placers (sands). In 1886 the South African Witwatersrand was discovered, the richest gold ore region in the world, where the production of gold was first placed on an industrial basis, on the rails of large-scale capitalist economy. Important technical innovations were soon introduced making it possible to mine gold-bearing ore at unprecedentedly low depths and to increase greatly the percentage of gold obtained from the ore. Ever since the end of the 19th century the fate of the gold-mining industry in the capitalist world has been indissolubly linked with southern Africa.

In 1886 less than a ton of gold was mined in southern Africa, and in 1898, 117 tons. After a sharp drop caused by the Boer War gold production began to rise again sharply. In the first decade of the 20th century southern Africa ranked first in world gold output. In 1913 the Union of South Africa (this state arose in 1910 as a British dominion) produced 274 tons, or 42 per cent of the world total. The USA came second with 134 tons.

The end of the 19th century witnessed the Klondike epic in Northern Canada and Alaska, immortalised by the pen of Jack London and the acting of Charlie Chaplin. This, however, had little influence on the main trends of gold production in the 20th century.

Until then gold production differed from other branches of

capitalist industry. It was more like a game of chance than a normal enterprise with good accounting. Thousands of prospectors were ruined and a tiny handful grew fabulously rich. In southern Africa it was a different matter. The companies here knew very well how much they spent per ton of processed ore and per ounce of gold produced. They sought to obtain a profit from gold production that was no lower than the profits in other industries. To a certain extent they could manoeuvre, change the amount of ore processed and gold obtained depending on the dynamics of production costs.

Another important change took place: up to the 1930s the *price of gold* in pounds sterling and dollars had remained practically unchanged for *two centuries,* because it was determined by the stable gold content of these two currencies.

The devaluation of the pound in 1931 and of the dollar in 1934 meant a sharp drop in the gold content of these currencies and, consequently, a rise in the price of gold. In dollars it rose by 69 per cent, and in pounds (by the beginning of World War II) even more. At the same time under the influence of the world economic crisis the prices of other commodities and, consequently, the costs of the gold-mining industry, dropped.

A strange thing happened: at a time when most industries were suffering from the depression cutting down production, the gold companies were flourishing and reaping vast profits. In 1940 gold production in the capitalist world reached its highest peak—1,138 tons, including about 40 per cent in the Union of South Africa. Second place in gold output was now held by Canada, with the United States coming third.

The next three decades were difficult ones for the gold producers of the capitalist countries. The mobilisation of industry for war needs resulted in a sharp drop in production in the USA, Canada and even in southern Africa. Under the influence of inflation there was a considerable rise in costs, while the official dollar price of gold, fixed in 1934, remained unchanged until 1971. The mining of gold was becoming less and less profitable, many mines were permanently or temporarily closed down. The South African industry survived this period more easily: new rich deposits were discovered there, technological progress made it possible to cut costs, and African labour power was still far cheaper than that of white workers.

In 1945 gold output in the capitalist world dropped to 654 tons, of which more than half came from southern Africa. By 1962 production had overtaken the pre-war peak, and in the late 1960s and early 1970s it settled at an annual level of 1,250-1,300 tons, with the Republic of South Africa (as it became called in 1961) steadily providing about three-quarters of this total.

A new era began in 1968, when the free market price of gold was allowed to deviate from the fixed official price. It moved only upwards, of course. Until 1972 the differences were comparatively small, but then, under the influence of inflation, the universal rise in raw material prices and the worsening monetary crisis, the free price of gold rocketed.

The profits of the gold-mining companies also rocketed, but the production of gold *went down, instead of up*. The companies behaved as befits monopolists: they restricted production to keep prices high. Technically this is done by going over to poorer ores, which it was unprofitable to process at the former price. In 1975 gold output in the capitalist countries dropped to below one thousand tons for the first time in many years.

Table 1

Gold Production in the Capitalist Countries
(in tons)

Countries	1913	1940	1960	1970	1975	1978	1979	1980
Total	652	1,138	1,047	1,273	956	969	968	960
including:								
Republic of South Africa	274	437	665	1,000	713	706	703	675
Percentage of total	42.0	38.4	63.5	78.6	74.6	72.8	72.6	70.0
Canada	25	166	144	75	51	52	49	48
Brazil	—	5	6	9	13	22	25	35
USA	134	151	53	54	33	31	28	29
Philippines	1	35	13	19	16	19	20	23
Australia	69	51	34	20	16	20	19	18
Papua-New Guinea	—	—	—	1	18	23	20	15
Ghana (Gold Coast)	12	28	27	22	18	14	12	13
Zimbabwe (Rhodesia)	22	26	18	15	16	11	12	11

Sources: *Currencies of the World*, Finansy i statistika, Moscow, 1981; *Bank for International Settlements, 51st Annual Report*, Basel, 1981.

In 1980 some 960 tons were produced, including 675 tons (about 70 per cent) in the Republic of South Africa. Second came Canada, as before the war, third was Brazil, and fourth the USA. However, the absolute volume of output in both countries of North America was much lower than before World War II. The remainder was made up by many countries, the most important of which are listed in *Table 1.* They include such traditional gold-mining countries as Australia, Ghana and Zimbabwe, and also the recent newcomer, Papua-New Guinea. There is hardly any gold-mining industry in Europe. Nevertheless, in some countries a small amount of gold is produced from old mines dating back almost to Roman times and as a by-product in the production of non-ferrous metals. In recent years all the countries of Western Europe taken together produced 10-15 tons of gold annually. Sometimes it is a matter of national pride to have one's "own" gold. The wedding ring of Queen Elizabeth II was made of British gold, a trivial amount of which is mined in Wales.

The most recent major discoveries of geological reserves of gold were made in the 1950s in southern Africa. Since then there have been occasional sensational reports of new discoveries, but all of them have turned out to be false or greatly exaggerated. The tremendous increase in the price of gold in the seventies stimulated prospecting to some extent, but so far this has yielded no important results.

As we see, the output of gold in capitalist countries is significantly lower than the pre-war level. At the same time the total production of goods and services has increased several times over these forty years, and many important metals are now being produced in quantities that are dozens of times higher than the pre-war level. This situation clashes with the growing demand for gold, producing some important consequences for the gold market.

The question of gold production in China is of special interest. Just as in the West insufficiently substantiated reports are spread from time to time about vast oil reserves in the People's Republic of China, so we find equally dubious information about Chinese gold. The world shortage of both types of raw material is the favourable ground on which such reports fall. However, specialists consider it is most unlikely that production in China will increase the amount of gold in the world significantly in the foreseeable future. Still reports worthy of attention have appeared in the Western press about new discoveries of and prospecting for gold in China. It is stated that recently discovered deposits may produce more than 30 tons a year. China's gold reserves are estimated at 400 tons. The Chinese are placing great hopes on Tibet, where, they believe, significant deposits of placer (alluvial) and primary (lode) gold will eventually be found. The development of the gold-mining industry in the People's Republic of China has encountered great difficulties. The working of deposits requires powerful foreign technology for the purchase of which it lacks foreign currency. China's gold policy today is basically to discover as many deposits as possible and to leave their exploitation until "better times".

Labour and Capital

The ages of mankind's development differ in the methods and tools with which people work. Changes in the instruments and modes of production form the basis of the changes in the structure of society, in the economic relations between people. This is most evident in the production of gold as well. In Ancient Egypt gold was mined with the help of stone and bronze pickaxes and any kind of large-scale production could exist only on the basis of slave labour. In the 20th century gold is mined with the use of complex technology: large capitalist enterprises and wage-labour correspond to this form of production.

This law should not be understood too literally. Although the Californian and Alaskan prospectors of the 19th century lived under capitalism, when there were already steam engines and railways, the technology that they used differed little from that of the Egyptians and Romans, yet the scale of work of those times was not accessible to them: the slave-owners were able to concentrate large masses of forced labour and even to introduce a certain division of labour between workers with different skills.

For whole millennia gold was mined mainly by the forced labour of slaves and serfs. In the ancient Eastern despotisms, including Egypt, they were state slaves, and gold production was a kind of state monopoly. Temples and corporations of priests could also be owners of slaves and gold mines. The Roman mines in Spain were also worked by state slaves.

These predominant social forms of production were not "pure" ones, of course. Quiring maintains that already in the second millennium B.C. the placer gold in Nubia and Kush (now the Sudan) was mined by free prospectors, whose only obligation to the state (the pharaoh and priests) was payment of a gold tax. He quotes Strabo's account of a "gold rush" in the 2nd century B.C., when crowds of freemen from all over Italy sped to the lands of the Taurisci tribe in the Eastern Alps (now Austria) where gold had been discovered. This displeased the tribesmen and they managed to drive away the gold-thirsty newcomers. Subsequently the reverse usually happened: the newcomers killed and drove out the inhabitants of lands that were fortunate or unfortunate enough to possess gold deposits. Pliny writes that at about this time the Roman state rented out to private individuals plots of gold placer deposits along the banks of the River Po, and that no single tenant was to use more than 5,000 workers on his plot. We must assume that the tenants were slaveholders and that the gold was washed by their slaves.[27]

In the Ancient World the reserves of gold produced were hoarded in state treasuries and passed from one conqueror to another. When the Roman Empire fell the gold was pillaged and for the most part disappeared from economic circulation in the Middle Ages: it went to the East and was buried in troves during the troubled time of invasions and mass migrations. The feudal sovereigns of Europe were forced to begin accumulating gold almost "from scratch". For this they made use of ownership of the land and non-economic coercion of serfs. The renting of gold-bearing land to teams of free or semi-free prospectors, and later to capitalist entrepreneurs, became increasingly widespread.

The growth of towns and the urban bourgeoisie was connected with the mining of gold in Silesia (now Poland). Entrepreneurs could pay rent to landowners and obtain a good profit, because wages were extremely low. At the beginning of the 16th century a worker earned 16 pfennigs a day; in order to earn one golden ducat (3.42 grammes of pure gold) he had to work fifty days! This made possible the working of the sands in the upper tributaries of the Oder, which were poor in gold.[28]

Frederick Engels considered that the concentration of the mining of precious metals in Germany in the 15th and 16th centuries was of great importance for the country's economic development and placed it at the head of Europe economically in this period. This, in turn, prepared the socio-economic conditions in Germany for "the first bourgeois revolution in religious guise of the so-called Reformation".[29]

However, the scale of gold and silver production in Europe corresponded less and less to the needs of the growing economy. "The discovery of America," Engels wrote elsewhere, "was due to the thirst for gold which had previously driven the Portuguese to Africa (cf. Soetbeer's *Production of Precious Metals*), because the enormously extended European industry of the fourteenth and fifteenth centuries and the trade corresponding to it demanded more means of exchange than Germany, the great silver country from 1450 to 1550, could provide."[30]

In the latter half of the 16th century the mining of precious metals in Europe dropped sharply, because it could not compete with the cheap silver and gold that began to flow in large quantities from America. The world centres of precious metal mining moved to the New World, where it was based on overt or hypocritically concealed slavery. The Spanish and Portuguese conquerors and their descendants received from successive kings plots of land together with the natives inhabiting them, who were essentially turned into slaves. The monstrous exploitation of the American Indians at the gold fields and silver mines became the main cause of the extinction of whole peoples in South and Central America. The importing of slaves from Africa began in the 16th century; many of them were sent to the mines. According to tradition the kings received one-fifth of all metal produced. This was a sort of tax in money and in kind: in kind because it was paid in the product produced, and in money because the product was money. It was not until the 18th, and particularly the 19th century, that capitalist relations began to develop in the mining of precious metals. However, by this time the share of South and Central America in the world output of precious metals had begun to drop.

Gold from America laid the foundation of the wealth of many noble feudal families in Spain and Portugal, but for the most part it was hoarded not in the gentlefolk's nests of Estremadura, but in the vaults of London banks: cloth mills and iron-making factories provided their owners with richer profits than the American mines. It was the flow of gold from the Spanish and Portuguese possessions that enabled Great Britain to adopt the gold standard virtually in the 18th century.

In Russia the state played the decisive role in gold production from the very outset. Up to the beginning of the 19th century it was officially a state monopoly. A decree of 1812 permitted private entrepreneurs to mine gold, but they were obliged to sell it to the state at a fixed price. Even then most of the metal was obtained at state enterprises. These enterprises infrequently made use of hired labour. The state gold fields and mines were often a place of hard labour, exile, imprisonment and punishment.

Hired labour was also rare at private gold fields and mines before the emancipation of the serfs in 1861. They were worked by serfs "attached" to the fields and mines or serfs on quit-rent from other places who had come to get seasonal work. It was only after the emancipation that capitalist relations began to develop slowly and with difficulty in the mining of precious metals, but right up to 1917 they were combined with many vestiges from the feudal-serf-owning times. The working and living conditions of workers in the gold-producing industry were extremely hard. Even at the beginning of the 20th century the working day was usually more than twelve hours long. Wages were often paid in coupons which could only be used to buy poor-quality and expensive goods at the mine shops.

In the works of D. N. Mamin-Sibiriak we find a vivid picture of the collapse of this military-bureaucratic and feudal system in the gold production of the Urals and the formation of the new, capitalist system, which contained many vestiges of serfdom. The novel *Gold* (1892) begins with the popular saying: we wash gold alone, but our voices groan. In the old days they groaned because of the slavery, the back-breaking work, the cold and hunger, and the flogging.[31]

Mamin-Sibiriak describes the work and life of these people with great insight. He also portrays buyers of stolen gold, whose activities incidentally make statistics of gold production in Russia less reliable. According to the official figures, from the beginning of the industrial production of gold to the overthrow of tsarism 2,754 tons of gold was produced in Russia. G. V. Foss, the author of a special work on gold, has grounds for assuming that the figure of 3,000-3,100 tons is closer to reality.[32]

In the endless expanses of Siberia, the Altai and the Far East gold was sought for, and occasionally found and worked, by groups of prospectors whose lives depended on luck. The economic status of these prospectors was similar to that of gold prospectors in California, Australia and the Klondike in the early years of working these placer deposits. In political economy terms it can be described as small-commodity production, that is, production where there is no capitalist and no wage labourer, where a person is working with his own tools and at his own risk. This was possible in the above-mentioned regions also because wild, unsettled lands usually belonged not to private individuals but to the state.

But capital lay in wait for the prospectors in the form of buyers of metal, sellers of simple, but nevertheless vital instruments, and traders in all manner of goods, particularly alcoholic beverages. Gold was bought up cheaply and goods sold at triple the price. And the actual mining did not remain in the hands of the pioneers for long. Very soon the plots of gold-bearing land were bought up by rich entrepreneurs who followed on the heels of the prospectors and began mining on a capitalist basis. A classic example is the fate of the privilege to the gold-bearing plot obtained by a certain George Harrison, considered to be one of the first discoverers of the wealth of Witwatersrand. Having neither the wherewithal nor the opportunity to set up an enterprise, he sold it in 1886 for £10. Ten years later, after passing through several hands, it was worth about £50,000.[33]

From the very outset in South Africa there was no place for prospectors with a spade and pan. Working the gold-bearing conglomerates required capital, technology and organisation. At first joint-stock companies which obtained money from local and foreign (particularly English) investors sprouted like mushrooms. In 1889, three years after the discovery of the gold, there were 450 companies. But even then the enterprises backed by the big capitalists towered above the small fry. Subsequently, independent companies sprang up by the dozen under favourable conditions and also collapsed by the dozen under the blows of crashes on the stock exchange and lack of success in the search for rich deposits. There was also a great deal of sharp practice around all this: the knights of profit warmed their hands at the expense of the small shareholders, and at the very lowest tier of this pyramid there were the rightless, poor, illiterate African miners deprived of all rights. At the beginning of the 20th century tens of thousands of Chinese were brought to the mines.

The process of the *concentration of production and capital* was taking place in the gold-mining industry: an increasing percentage of production was done by a few large enterprises and an increasingly narrow group of financial magnates controlled these enterprises. In all, seven large companies control South Africa's gold-mining industry now. This structure has existed without any great changes for several decades: the youngest of the "sisters" was born in 1933.

Each of the companies is a highly complex organisation. We usually call this type of structure of large capital a *concern*. They often refer to themselves as *groups*. Each concern is built on the pyramid principle: the highest tier controls the lower ones. This control is exercised through the holding of a controlling interest in the share capital. A concern is a system of joint-stock companies with a fixed hierarchy of authority and control. Another most important feature of the concern is its diversified structure: as well as gold mines, each of them includes enterprises in other branches of industry, trade, the service sector, and also financial companies.

The most powerful South African concern is the Anglo-American Corporation of South Africa Limited. It was founded in 1917 by the diamond magnate Ernest Oppenheimer and is still controlled by this family. Today the Anglo-American Corporation is essentially a multinational corporation with its centre in Johannesburg, but with extensive interests in the industry of many capitalist countries.

The Free State Geduld Mine mentioned above is part of the Anglo-American group. What does this mean?

Formally the mine is owned and run by an independent joint-stock company known as the Free State Geduld Mine. The controlling block of shares belongs to the holding company Orange Free State Investment Trust Limited. The exact size of this controlling interest is not known. But according to usual practice in the gold-producing industry of South Africa 25-30 per cent of the shares is enough for a controlling interest, sometimes even as little as 10 per cent. The remaining shares are spread among the mass of shareholders. They may include both big fish, other concerns for example, and small investors from all parts of the world.[34]

A holding company is a purely financial organisation. It produces nothing, but merely holds shares and collects dividends on them. The assets of the Orange Free State Investment Trust are invested almost entirely in shares in the mines situated on the territory of the Orange Free State. But it is a subsidiary of the head concern Anglo-American, which is itself nothing but a gigantic holding company.

As well as holding controlling interests, the whole system is cemented also by *interlocking directorates*—the occupying of the top administrative posts by one and the same people. Harry Oppenheimer (son of the empire's founder) is Chairman of the Board of Directors of the head concern, but he is also a member of the Board of Directors of the holding company Orange Free State. And that is run by a man who is a member of the Board of Directors of Anglo-American.

The management of each mine administers the mine, but its powers are limited. Questions of new capital investment, financing, technological policy, recruitment of labour and marketing are all decided at the centre, in the headquarters of the holding company and the whole concern.

In recent years the Anglo-American group accounts for about 40 per cent of the total output of gold in South Africa, leaving their nearest competitors far behind. But for the concern itself the importance of gold as a sphere of capital investment is tending to decrease. No new rich gold fields have been found in South Africa over the last twenty-five years and the profitability of gold mining was dropping right up to the beginning of the 1970s. Timothy Green says that Harry Oppenheimer told him in 1967: "You can't sit with your money waiting for a new gold field."[35] In order to live a firm must grow. And the Anglo-American concern is growing rapidly. Its daughter, granddaughter and great-granddaughter companies are mining diamonds, copper, uranium and coal in South Africa and other countries, manufacturing steel and paper, chemical goods and beer, buying and selling plots of land, and granting bank loans. The other large companies have pursued the same policy in recent years.

The three largest concerns (Anglo-American, Gold Fields and Union Corporation) together own 27 gold mines which produce 70 per cent of South Africa's total gold output. But each of them, particularly Anglo-American, has an interest in the other concerns, or in the holding companies and mines controlled by them. All seven concerns are interconnected in the most complex way by the system of mutual holding of controlling interests and directorial posts. Essentially the concerns form a kind of cartel within the framework of which competition is severely restricted. Real competition exists only in the search for new deposits and gaining control over them. The concerns do their best to keep their work and achievements in this sphere secret. If a rich lode is discovered, the concern tries to buy up quickly and secretly the land under which it is likely to run.

The mining-financial concerns that have grown up on gold are the backbone of young, aggressive South African monopoly capital, the economic foundation of a state which practises racist oppression in its most blatant forms. An important role has always been played in the development of South Africa by foreign capital, British in particular. But the proportion of foreign shareholders has been gradually dropping. The modern South African concerns are closely linked with American and British capital, but they are an independent and major force. It is hard to call Oppenheimer and the other owners of the gold-mining industry in South Africa representatives of South African capital alone. Basically it is cosmopolitan capital.

Both in South Africa and in other countries it is the big concerns that are engaged in gold production. Only a tiny fraction of total output comes from the enterprises of small firms and even less from "free prospectors". In addition the gold-mining industry is growing more and more closely involved with other branches, becoming only one of the spheres of the concerns' activities.

IV. WHERE THE GOLD GOES

The production of a commodity is only the beginning of its life's path. After the commodity has been produced, it must be realised (sold). Often it passes through several acts of realisation, for example, the producer sells it to a wholesaler and the wholesaler to a retailer. Eventually the commodity reaches the consumer and is consumed in accordance with its natural form: food is eaten up, clothing is worn out, and machinery is gradually used up in production or in the home.

Gold passes through all the same stages as other commodities. But its special social functions leave their mark on all these stages.

We have already seen how this manifests itself in production. Let us now examine how gold is sold and "consumed". The word is in inverted commas because at this stage the difference between gold and all other commodities is particularly striking. For decades the life of gold was as brief as that of moths. It saw the light of day in a South African mine only to disappear into the vaults of state treasuries. Its consumption consisted in lying there, massive and motionless. At first it was thought that this gold served as reliable collateral for paper-credit money. Later the formal connection of the gold reserve with the domestic money supply was severed in practically all countries, and its real significance in this role was reduced to nothing. The functions of central gold reserves were concentrated in the foreign economic sphere, and gold began to be regarded as the last line of monetary defence, as a guarantee of a country's foreign solvency. This function of gold reserves survives today, but some important and contradictory changes are taking place in it. On the one hand, the formal demonetisation of gold in the system of the International Monetary Fund (1970s) would seem to lessen the importance of gold reserves. On the other hand, the repeated rises in the price of gold are increasing the real value and effectiveness of gold reserves.

In recent years freshly produced gold has not gone to replenish the central reserves. The nature of its "consumption" has, thus, changed significantly. This "consumption" reflects major trends in the development of capitalism at the end of the 20th century.

Where It Is Now

In economics and book-keeping there are two main ways of measuring quantities: *at* a certain date ("stock") and *over* a certain period, usually a year ("flow"). In the first case we have, as it were, a photograph of the financial state of affairs (of an enterprise, company, country, etc.) at a given moment, and in the second a picture of the flow of assets: where they came from and where they have gone.

This twofold approach can be conveniently used for an analysis of the movement of gold in the capitalist world. Let us first try to answer the question of where is the accumulated gold (stock).

As we know, there is now no gold money in circulation. The nearest thing to gold seen *as money* are the central reserves belonging to governments or central banks. In economic literature this gold is sometimes called monetary gold. It is by no means the same as gold *coin*. On the contrary, today monetary gold takes the form almost exclusively of standard ingots of great fineness.

It was not always like this. In 1913 almost half the total of monetary gold (5,900 tons out of 13,200 tons) was in circulation in the form of coins. More than half (7,300 tons) was concentrated in central reserves.[1] There is reason to believe that a considerable part of these reserves also took the form of coins. Already in the course of World War I combatants and even non-combatants used all the methods at their disposal to take gold coins out of circulation. By the late 1920s the USA was the only place where there was still a small amount of gold coins in circulation.

Russia's gold reserve, which was removed from Kazan by whiteguards in 1918, taken eastwards, and returned to Soviet power through the heroism of the Red Army and the Siberian partisans, consisted almost entirely of coins. When the gold was received by Soviet bodies in Irkutsk in March 1920 there were 619 poods (more than 10 tons) of ingots and 20,823 poods (341 tons) of coins of various mintings in the carriages of the "gold train". It was valued at 409.6 million old gold roubles. What is more, during the train's two years of wanderings Kolchak's men stole about 242 million roubles and, in all probability, this gold also was in the form of coins, not ingots.[2]

And here is a fact from the more recent past. In 1963 the Uruguayan Central Bank requested two foreign banks specialising in commercial operations with gold to exchange coins worth 60-80 million dollars from its gold reserve, which then weighed 50-70 tons, for ingots. In the course of the operation it transpired that many of the coins were of numismatic value. This fact attracted attention because previously Uruguay's gold reserve had been in an unusual, old-fashioned form.[3]

According to the statistics published regularly by the International Monetary Fund central reserves of monetary gold totalled in 1979 35,200 tons. This is five times more than in 1913, but slightly less than in 1967, when the gold reserves of the capitalist world reached their maximum (37,700 tons) (see *Table 2*). The percentage of central reserves in the total stock of hoarded gold has dropped significantly in recent years.

The largest holder of gold is the government of the United States although its share has dropped significantly compared with the period after World War II and is continuing to drop. At the end of 1979 the gold reserves of the USA stood at 8,230 tons, which represents 28.5 per cent of all central reserves, not counting those of international organisations. The second largest holder is the Federal Republic of Germany, which hoarded gold in the fifties and sixties thanks to its stable balance of payments. The list of countries with large gold reserves includes a number of other industrially developed countries. This is the well-known "Club of the Rich". All the industrial countries taken together, at the end of 1979, accounted for 84.8 per cent of world central reserves. Of the developing countries only a few oil-producing ones have any significant gold reserves. But even the largest gold reserve in this group, that of Venezuela, was only 356 tons at the end of 1979. Most developing countries have a trivial amount of gold. A large holder of gold is the International Monetary Fund which, in accordance with its original Articles of Agreement, required each member country to subscribe a quota partially payable in gold. But in 1976-1980 the Fund's gold reserve dropped by about a third when, in accordance with the Kingston Agreement of 1976 on the reform of the international monetary system, it returned one-sixth of its reserves to member countries and sold another sixth on the free market. We shall return later to the question of the IMF's gold.

In 1979 in connection with the creation of the European Monetary System uniting the Common Market countries, these countries each transferred 20 per cent of their gold reserves to the European Monetary Co-operation Fund, and received in return a claim upon this fund in the form of special European Currency Units (ECU). As a result the European Monetary Co-operation Fund also became the holder of a considerable gold reserve. A certain amount of gold is also held by the Bank for International Settlements in Basle, whose members are the central banks of many countries all over the world.

At the same time as the physical amount of gold in central reserves was dropping slightly, its monetary value had risen dramatically by comparison with 1967, when the official price of 35 dollars an ounce was still operating and the free market price did not differ from it significantly. At the market price of late 1979 the gold reserves of all countries and organisations added up to the colossal sum of 579 billion dollars. They exceeded the total sum of all other international monetary reserves at the end of 1979, which, according to IMF statistics, was 356 billion dollars. This sum includes the foreign exchange assets of central banks and governments, unconditional rights to the credit resources of the IMF and Special Drawing Rights (SDR).

The world's biggest depository of gold is in New York, deep down in the rocky ground of Manhattan. It is the vaults of the Federal Reserve Bank of New York, the most important of the twelve banks that form the Federal Reserve System which in the USA performs the functions of a central bank. In addition to part of the USA's gold this gigantic store house contains in full or in part the gold reserves of more than seventy countries of the capitalist world and of the International Monetary Fund. In August 1972 about 13,000 tons of the metal were stored there. Most West European states with the biggest gold reserves store a large part of them in New York. The exception is France which, in accordance with tradition and the policy of its government over the last two decades, holds its gold reserve on its own territory.

Behind a 90-ton steel hermetic door, at a depth of 85 feet (about 30 metres) lie these riches, beside which the real and legendary troves of all times pale in comparison. The gold is stored in 120 steel safes of different sizes. Each safe has three locks, the keys to which are kept by different officials of the

Table 2

Central Gold Reserves of Capitalist Countries
(at the end of the year)

	1967		1979	
	tons	billion dollars*	tons	billion dollars**
All countries	35,100	39.4	28,925	476.1
Industrially developed countries	31,085	35.0	24,542	404.0
USA	10,721	12.1	8,230	135.5
Federal Republic of Germany	3,757	4.2	2,963	48.8
Switzerland	2,745	3.1	2,590	42.6
France	4,651	5.2	2,548	41.9
Italy	2,132	2.4	2,075	34.2
Netherlands	1,524	1.7	1,368	22.5
Belgium	1,314	1.5	1,064	17.5
Japan	300	0.3	754	12.4
Austria	623	0.7	657	10.8
Great Britain	1,147	1.3	568	9.3
Oil-producing developing countries	993	1.1	1,137	18.7
Non-oil-producing developing countries	3,022	4.1	3,174	52.2
International Monetary Fund	2,382	2.7	3,322	54.7
European Monetary Co-operation Fund	—	—	2,653	43.7
Other international organisations	187	0.2	268	4.4
Total:	37,670	42.3	35,168	578.9

* At the official price of 35 dollars per troy ounce (31.1 grammes).
** At the London market price for the end of 1979—512 dollars per troy ounce.
Sources: *International Financial Statistics Yearbook 1979*. IMF, Washington, 1979. *International Financial Statistics, October 1980*, IMF, Washington, 1980.

bank. A holder of gold may have one or more safes depending on how wealthy he is. The largest safe, says Green, from whom this description is borrowed, contained 110,000 standard bars (about 1,400 tons). In accordance with banking tradition, the officials refused to divulge the name of the owner of the contents.[4]

The sale (and purchase) of gold by one country to another is done by means of transferring bars with the help of ordinary carriers and conveyor belts from one safe to another. If the amount of gold sold is equal to or more than the contents of the safe, the number of the safe is simply changed and the gold is transferred to a new owner by means of a simple paper operation.

If the gold is sold by the USA and its small working balance in the bank vaults is not sufficient, metal is brought from the store house of the US Assay Office, which is in a different part of New York. This store house is a kind of branch of the famous Fort Knox in Kentucky.

In 1974, when there was much discussion about gold in connection with the repeal of the law prohibiting private ownership of monetary gold by American citizens, the US press began to spread rumours that Fort Knox did not contain the amount of gold stated in Treasury reports. Vigilant Congressmen demanded a reply from the Treasury. On 23 September 1974 a group of ten members of Congress and several dozen journalists were admitted for the first time to this mysterious "house of the yellow devil" which contained about 4,500 tons of gold.

The next day in all the leading newspapers the newsmen described with malicious delight the grey, squat, outwardly unremarkable building, the cordons of guards, the checking of documents and pockets, and the secret numbered locks. The 22-ton steel door finally opened and the overawed visitors entered the first of the thirteen chambers in which the gold is stored. The doors are sealed with the personal seals of the responsible Treasury officials. The visitors were shown only one of these chambers, in which there was about 500 tons of gold. No one had entered it since 1968. The piles of bars gleamed dully. Congressman John H. Rousselot of California, who had been particularly zealous in demanding a check, took one look at the gold and said: "I think it's there." The rest agreed with him, although no one offered to show them that all the 367,500 bars stored in Fort Knox were actually there and had the standard weight and fineness. Such a check was intended, but it would have taken many weeks.

Fort Knox is rarely opened—only in order to remove or add a few hundred tons. When this happens a special convoy of armoured cars with a reinforced guard sets out from there to New York or vice versa.[5]

According to the latest statistics there is more than 20,000 tons of monetary gold on the territory of the USA, which represents 56 per cent of the central reserves of the capitalist world.

It is highly likely that the world's third largest gold depository is in the centre of Paris, in the vaults of the Banque de France. René Sédillot, a leading specialist on the history and problems of gold, describes the armoured subterranean store chambers with their heavy doors, and the main columned hall of this Temple of Gold which reminded the author of the temple of Memphis. "Did you know, Parisians, that this rock stretches beneath the sand, marl and stones of your earth and that France hides its gold in its bosom? The subterranean universe of the Banque, which is situated under the new buildings near the Palais-Royal, is connected with the surface only by four wells which are guarded by eight armoured towers. The gold is protected both from human lusts and from the tumult of wars."[6]

To avoid the tumult of World War II, the French sensibly packed off their gold to their overseas colonies, thanks to which it did not fall into Nazi hands. The fate of the French gold reserve reflects the history of the Third, Fourth and Fifth Republics. In the early 1930s it reached 5,000 tons, but by the beginning of the war it had dropped considerably, and in the difficult post-war years it fell to a few hundred tons. But under de Gaulle (1958-1969) France again accumulated a large gold reserve.

Over the last fifteen years peace has mostly reigned in all these citadels of gold. In 1968 the main capitalist countries agreed not to buy or sell gold on the free market, and the total volume of central reserves was thus frozen. In 1971 the US closed the "gold window" by unilaterally refusing to exchange dollars belonging to foreign central banks for gold. Settlements between countries in gold practically ceased, and the gold reserve of *each* country was frozen, so to say. In the following chapters we shall describe how this "dead" gold has begun to come to life and what this presages.

The remaining gold of the capitalist world is in private hands (firms, banks and private individuals). For obvious reasons these figures are only the roughest estimates. It is believed that *about 26,000 tons is in private hoards.* This gold is a specific form of investment of money capital and savings. Inflation and socio-political instability have in recent decades and years enhanced the attraction of gold for different strata of the bourgeoisie and other groups of the population.

The gold in private hoards takes varying forms. Among small hoarders gold coins are particularly popular, hence the "coin boom" of recent years. Large investors now frequently store metal (usually in banks) in the form of standard monetary bars or the right to demand them.

It is thought that the largest amount of gold is hoarded in France, where this custom is widespread among the petty and middle bourgeoisie, the urban middle strata and the peasantry. Specialists usually agree on the figure of 5,000-6,000 tons. About the same figure is quoted for India, which may seem strange at first glance. But the primitive gold ornaments and coins in Indian peasant families are a form of insurance against a bad harvest or other disaster. Among the middle and big bourgeoisie in India (and in other countries of the same type) it is customary to parade one's gold in order to demonstrate wealth, respectability and stability. As Gunnar Myrdal (the Swedish economist and Nobel Prize winner) once remarked, "It is in the southern, not the northern, part of the American hemisphere [he had in mind Latin America— *A.A.*] that it may happen that one is invited to eat on gold plates."[7]

Up to 1975 citizens of the United States were prohibited by law from owning gold (with the exception of articles of jewelry and numismatic valuables). Nevertheless, according to some estimates, at the time when this ban was raised rich Americans had circumvented the law and were holding almost exclusively abroad, particularly in Swiss banks, from 4,000 to 4,500 tons of gold.[8]

After 1975 a stream of gold poured into the USA for hoarding. Individual capitalists, firms and financial institutions there soon developed a tendency to hold part of their money capital in gold permanently. The tremendous rise in speculative dealings with gold demanded an increase in commercial firms' working balances. In three years alone (1976-1978) experts estimate the growth of investments in gold in the USA at 515 tons, and this process was accelerating each year.

Traditionally considerable amounts of gold are hoarded in the countries of the Middle East and Southeast Asia. The flow of gold into some of these countries has increased in recent years in connection with growing incomes from the sale of oil. Many different factors influence the absorbing of gold by private hoarding. For example, in 1978-1979 there was a sudden explosion in the demand for gold in Taiwan in connection with the United States' recognition of the Peking government and the consequent uncertainty in the status of Taiwan.

The third form which hoarded gold reserves take is that of *articles of jewelry, dental work, parts of apparatus,* etc. The dividing line between jewelry and hoarding is blurred and conditional. In many cases, particularly in Eastern countries, articles of jewelry perform the function of storing stable value. Part of the gold used in electronic and other apparatus is lost forever and disappears from economic circulation, but a considerable part of it returns as secondary raw material. Estimates of gold existing in these forms are extremely approximate and range between 20,000 and 25,000 tons.

We can assume that in all there are about 80,000 to 85,000 tons of gold accumulated in the capitalist countries in various forms. This figure corresponds with the estimates of total output and permanent losses of gold quoted above. The existence of this reserve of economically active gold has an important influence on the gold market, endowing it with features not found on the market of any other commodity.

How It Is Bought and Sold

We live in a rapidly changing world. Twenty years ago only academics and theoreticians wrote about the international demonetisation of gold, that is, about depriving it officially of monetary functions, whereas today this is part of the Articles of Agreement of the International Monetary Fund. Of course, the juridical act does not solve the economic problem, but it does reflect important real changes in the status of gold. These changes are particularly evident on the gold market. To a considerable extent it has lost the features of a specific market of a monetary metal and acquired those of a raw material, commodity market. The same thing is happening to gold that has happened to silver.

Formerly governments and state central banks bought up almost all freshly produced gold and controlled its subsequent movement. They sold it to one another when they needed to supplement their monetary reserves, and also to private firms which used it for industrial purposes. In 1968 governments and central banks withdrew from the gold market. If they occasionally appear on it today, it is only "on an equal footing", to buy or sell not at a fixed monetary price, but at the market "commodity" price.

A somewhat confused question may arise. We know that one of the most important trends of modern capitalism is the strengthening of the economic role of the state: it is increasingly producing, buying and selling, regulating and controlling. But something like the reverse would seem to be happening to gold. In fact there is no contradiction here. The point is that previously gold was the foundation of national monetary systems and of the international monetary system. These are vitally important spheres for the bourgeois state. But with the limiting of the monetary functions of gold the need for state intervention also decreased. The centre of gravity of this intervention is now shifting to the more essential spheres of money and credit, balance of payments, and exchange rates.

Of course, no one can say that the monetary role of gold on the world market is finished, or that gold has become only one more type of metallic raw material and nothing else. We shall return to this important question in later chapters. Here we would only point out that *at the present stage* the gold market is in the hands of large *private* capital.

Under the gold standard gold was money. But selling means exchanging a commodity for money, and the price of a commodity is its value in terms of money. Therefore expressions of the *buying and selling* and *price* of gold were at that time merely conventional phrases, or, as Marx put it, irrational concepts.

The price of gold was simply the gold content of the monetary unit expressed in a different way. People said the price of gold in the USA was 20.67 dollars per troy ounce. What they meant was: the gold content of the dollar (fixed by law in 1900) was equal to approximately 0.048 ounces, or 1.505 grammes. In selling gold, the seller was simply exchanging, for example, an ounce of ingot gold for 20 dollars 67 cents in gold coins or banknotes which were always freely convertible for gold. This was, so to say, merely a change of form.

But when the dollar ceased to be convertible into gold, it was possible to sell gold at the new official price of 35 dollars an ounce, but not to buy it. The seller of gold, an American gold industrialist, for example, did not change it into a different form now, but exchanged it for something radically different, paper dollars. This gave the operation the features of a real sale.

After World War II when the prices of all commodities rose sharply, it became unprofitable to sell gold to the government of the USA at 35 dollars an ounce, and everyone who could began to avoid doing so, preferring to sell it unofficially at a higher price. A *free gold market* arose with its own prices, which constantly fluctuated like the prices of many other commodities in the capitalist economy.

In the 1960s the governments of the USA and their allies still thought it necessary to support the fiction of a stable official price of gold, that is, the stability of the gold content of the dollar, the fictitious equality between the dollar and gold. To this end in 1961 the so-called *Gold Pool* was set up, which included the USA and the main West European countries. The aim of the Pool was to maintain the price of gold on the free market at the official level. The method of attaining this aim was to regulate the market by the buying and selling of gold on the London market where gold operations were concentrated. If the free price dropped, the members of the Pool bought up the gold surplus and divided it between them in accordance with a previously agreed formula. If it rose, they sold gold and divided the costs according to the same formula. The USA bore half of these costs and after France left the Pool in the summer of 1967 its share grew even larger.

This system collapsed in 1967-1968. In the USA and the other capitalist countries inflation was growing and the economic situation deteriorating. In November 1967 Britain, the weakest link in the system, devalued the pound sterling, that is, reduced its gold content and exchange rate in relation to the dollar and other currencies. This was followed by panic and feverish speculation: big and not so big capitalists began to buy up gold on a large scale, expecting that its official dollar price would go up. The Gold Pool kept dumping hundreds of tons of gold on the market, but the demand was insatiable. Private speculators wanted gold in its natural form, not in the form of New York bank receipts. So the government of the USA even had to use military aircraft and an American military base near London to convey the precious cargo quickly enough.

In a futile attempt to maintain the equality between the dollar and gold the members of the Gold Pool spent about 2,500 tons of gold between November 1967 and March 1968. In one day, 13 March 1968, 175 tons was sold in London, where nearly all the dealing took place, and the next day 225 tons.[9] The speculation had reached its height. The speculators ran no risk of losing because the price of gold could only go up, not drop. The USA's partners in the Gold Pool rebelled, and the blood-letting from the American gold reserve was also felt. As usually happens in the financial sphere, one day the players in the game observed its rules and announced that they would defend the official price of gold (and, consequently, the parity of the dollar) down to the last bar in the bank vaults, and the very next day they announced the dissolution of the Gold Pool and the end of all attempts to regulate the market. The so-called two-tier market and two-tier price of gold arose. The central banks of the main capitalist countries undertook not to buy or sell gold on the free market, leaving it entirely at the disposal of the private sector. The free price broke away from the official price of gold and rose considerably. Insofar as the speculators had bought a vast amount of gold and were now realising the profit by selling it, the free market price in 1968-1970 was not very different from the official price (not more than 10-15 per cent). Only in 1971 did it begin to rise more sharply.

The price of gold is measured in paper-credit money and is the same economic reality as the prices of all other commodities. Only through these prices is the ratio of gold to other commodities established. If, say, an ounce of gold costs 100 dollars and a ton of wheat costs 100 dollars, from this alone we can conclude the price of wheat expressed in gold (1 ton of wheat = 1 ounce of gold).

Gold comes onto the world capitalist market from three main sources: from the current output of the capitalist and developing countries, from accumulated central reserves, and from other sources, including the sales by socialist countries *(Table 3)*. The annual total of gold that reaches the market in these ways has been 1,300-1,800 tons in recent years. Current output changes relatively slowly and does not fluctuate much from year to year, whereas the supply from the other two sources may vary considerably. The largest seller of gold on the world market is invariably the Reserve Bank of South Africa, which has a monopoly on the sale of gold produced in South Africa.

This amount represents the inflow of gold onto the market. But one and the same ingot can, of course, circulate several times before finding its "consumer". Therefore the annual turnover of markets may be several times higher than the amount that comes on the market.

Only a very small part of the gold goes straight from the producers to the consumers, the industrial firms that use it as a raw material. Traditionally nearly all gold first passes through the hands of middlemen, usually called dealers.

Gold *dealers* are the financial aristocracy, the elite of the business world. Their names are associated with such important elements in the gold business as financial power, respectability and secrecy. This is a traditional club which no newcomer may enter. The number of dealer firms that count on the market does not exceed twenty, and they are all either big banks or very closely connected with them. In order to carry out operations with gold you need big money.

Before World War II the world centre of the gold trade was London or, rather, its financial part, the City. In spite of its losses, London is still of great importance today. According to the latest estimates, about one half of South African gold is sold through London dealers, and the other half of this tasty pie has in recent years gone to the Swiss banks, the notorious "gnomes of Zurich". Since 1975 some South African gold has been sold in the USA through American firms.

The list of London dealers is headed by the House of Rothschild, founded in 1804. Its official title is N. M. Rothschild & Sons Ltd. Many a book has been written about the Rothschild family and financial empire. Their name is wreathed in many a legend: the Rothschilds stand for two centuries of large and daring financial operations, fantastic wealth, considerable overt and far greater covert power. From the founder of the dynasty, who engaged in small-scale money-changing in Germany in the second half of the 18th century, right up to the present day, the name of Rothschild has been associated with gold. The Rothschilds invested capital in the Witwatersrand gold mines and have been receiving dividends from them for almost a century now.

At the Rothschild bank, in a room furnished in old-fashioned style and decorated with the portraits of European monarchs, clients of the Rothschilds, the traditional procedure of "fixing" takes place every day, i.e., fixing the approximate price of gold on the basis of which the actual dealings will take place. This price is instantly telexed and telephoned to New York, Zurich, Toronto, Singapore and other centres of gold trading and serves as the basis for price formation on all markets.

The "Big Three" Swiss banks have their own considerable advantages: the traditional neutrality of the Alpine republic; the money that flows in from all over the world; and the equally traditional banking secrecy which extends to all clients including fugitive dictators and criminals. In Switzerland there are no restrictions on the ownership, buying, selling, import and export of gold and no taxes on gold dealings. In contrast to the London market which is predominantly a wholesale one, the respectable Zurich gnomes do not balk at connections with smugglers and corrupt officials and have created for themselves an extensive network of retail clients in France and the countries of the Arab East and South and Southeast Asia. They supply gold there in small bars, in coin.

In the Middle East the largest gold market was in Beirut, but recently business has shifted to Kuwait and Dubai. In the Far East the centre of gold trading has long been Hong Kong and Macao. In recent years the Singapore market has had a considerable turnover. Accounts of the gold business in Asia sometimes read just like detective stories and thrillers: intrigues, bribery, smuggling, mysterious murders, cut-throat competition.

The American monopolies could not and did not stand aside from the gold business. Until recently their activity was somewhat restrained by the fact that citizens of the USA did not have the legal right to own gold. Today these restrictions, introduced in the 1930s, have been abolished. In New York and Chicago gold markets are operating which threaten to take over from London and Zurich. US industry has always been the world's largest consumer of gold, and in recent years the Americans overtook the French as private investors and hoarders.

The large dealer firm Philipp Brothers operates within the framework of the Engelhard Minerals and Chemicals Corporation. The name of Engelhard means almost as much in the gold business as those of Oppenheimer and Rothschild. Another branch of the concern owns the largest gold-processing enterprises in the USA. They produce semi-finished products that are purchased by the jewelry, electronics and other industries, where the gold finds its final use.

How is the current, annual gold flow divided between the main spheres of final use (consumption)?

Although in principle the purchase of gold on the free market by central banks and international finance organisations is possible and is practised from time to time, on the whole since 1967 this sphere, as we know, has not only not been devouring the metal, but has even regurgitated some of it (see *Table 3*). Some say this is a sign of the demonetisation of gold and the rational use of dead treasure, others that it is a silly and dangerous experiment. We shall return to this question later, and shall limit ourselves here to some facts and figures.

In the period 1968 to 1976 an annual average of 1,321 tons of gold came onto the market. It flowed off in two directions: the sphere of industrial use and the sphere of private hoarding. Industry consumed an average of 1,014 tons (77 per cent), including jewelry-making 791 tons (60 per cent), electronics—93 tons (7 per cent), and dental work and dentures—67 tons (5 per cent). These are the big three in the industrial use of gold. The remaining 5 per cent went to all the other spheres.

Private hoarding accounted for an annual average of 307 tons (23 per cent) of all the gold coming onto the market. This amount was divided almost equally between the hoarding of bars and the hoarding of coins.[10] These figures might seem to underestimate the amount of gold absorbed by private hoarding. The point is that a considerable part of gold articles, particularly in India and other Eastern countries, is not accumulation of special articles of consumption, but hoarding of the embodied value, of a specific commodity retaining certain features of money. In the Middle East and India ornaments of very fine gold are popular. In their price the metal (raw material) accounts for up to 80 per cent and the work of the jeweller for only about 20 per cent. It is easy to turn them back to pure metal again with a comparatively moderate loss.

Obviously a considerable amount of the jewelry consumption of gold is in fact a form of hoarding. The above-quoted figures

Table 3

Supply and Demand for Gold on the World Capitalist Market

	in tons						
	1974	1975	1976	1977	1978	1979	1980
Supply							
New output	1,006	954	970	972	979	961	943
Sale from central reserves	20	9	58	269	362	544	—230
Other sources	220	149	412	401	410	199	90
Total	1,246	1,112	1,440	1,642	1,751	1,704	803
Demand							
Jewelry industry	225	523	935	1,003	1,007	737	120
Other industries	216	188	216	223	252	255	207
Industrial demand	441	711	1,151	1,226	1,259	992	327
Investment and hoarding demand*	805	401	289	416	492	712	476
Total	1,246	1,112	1,440	1,642	1,751	1,704	803

* Bars, coins and medals used in the private non-industrial sector.

Source: *Gold 1981*, Consolidated Gold Fields, London, 1981.

are only estimates which do not claim to be highly accurate.

In the following years, when the pursuit of gold as a means of preserving value and an object of speculation increased sharply, and prices showed a clearly expressed tendency to rise, the absolute amount and share of hoard increased considerably. In 1977 of the 1,642 tons of gold that entered the markets of capitalist countries 416 tons (more than 25 per cent) was hoarded in the form of bars, coins and medals. There were no significant changes in the structure of the industrial use of the metal. In 1978 the flow of gold onto the markets rose to 1,751 tons, and 492 tons (more than 30 per cent) went to hoarding. The markets absorbed the sharply increased amount of gold coin, particularly South African Krugerrands, which we shall discuss below. It is typical that, in spite of rapidly rising prices, the use of gold in the jewelry industry did not drop in absolute terms. This is indirect proof that people were increasingly buying such articles for the purpose of investment. In 1979, according to the statistics, of the 1,704 tons of gold that entered the market 712 tons, i.e., more than 40 per cent, was hoarded. These figures reflect the growing instability of the gold market, the increase in speculative demand, and the rejection of paper money that was losing its value.

In 1977-1979 a large amount of gold, about 1,200 tons, was put on the market from central reserves (mainly the International Monetary Fund and the US government). This met a considerable amount of the demand and evidently prevented prices from rising even more sharply. But another factor appeared at this time—the buying of gold by certain countries for their central reserves. The fact that in 1978 Saudi Arabia bought 45 tons of the metal has attracted special attention. In 1980 net buying by central banks amounted to 230 tons. If this source of demand extends significantly in the future, it may have a considerable influence on the market and price of gold.

Smuggling

In the post-war period gold smuggling turned into a whole branch of business with its own methods, its own organisation and bosses, and its own profits and risks. Green, who has written a book on gold smuggling, believes that about half the gold that entered the market in the early seventies had passed through the hands of smugglers and filled their pockets. These hands and pockets belonged to different people: the "hands" are hired couriers who receive a very modest "wage" (the hiring and work of one of these couriers is described in Allan Sillitoe's novel *A Start in Life*); the pockets belong to the bosses of the international smuggling syndicates—capitalist organisations. "In fact," writes Green, "over the years a variety of extremely efficient international syndicates have grown up, whose turnover, if declared, might well place them in the world's top 500 companies."[11] These people, however, prefer for a number of reasons to remain in the shadows: their activity is connected with breaking the laws of many countries and their profits avoid taxation.

Smuggling has arisen and prospered because many countries prohibit the import, export, ownership, and buying and selling of gold. According to figures for 1970 there were only 10 countries and territories in the capitalist world where these "freedoms" existed: West Germany, Switzerland, Netherlands, Canada, Paraguay, Lebanon, Saudi Arabia, Kuwait, Dubai and Macao.[12] True, the list has increased since then. Among the major countries that have lifted their bans we can mention the United States, Great Britain and Japan. But there are still very many countries that retain bans and restrictions. Smuggling is changing its direction and forms, but certainly not disappearing.

Where gold operations are prohibited or restricted, its price is higher than on the free international markets of London or Zurich. When gold cost 35 dollars an ounce in London (this price was supported until 1968 by the above-mentioned Gold Pool), the price was 50-55 dollars in India. The difference per kilogram of gold was 500 dollars or more. And smuggling was considered profitable even when the difference was only 75-100 dollars per kilogram.[13]

We already know that the buying of gold by the public is often a sign not of wealth but, on the contrary, of poverty and economic chaos. This has been the position in Egypt in recent years. Western specialists estimate that in 1976 up to 25 tons of gold was purchased there (mainly in the form of primitive articles of jewelry). Its price was 20-30 dollars higher than the

6-437

international price which was 125 dollars an ounce on average that year. The surcharge here is relatively lower than in India, not because the syndicates were less greedy, but rather because the Egyptian government cannot (or will not) effectively combat smuggling.[14]

The general pattern is obvious: the weaker the control, the easier smuggling and the closer the domestic price to the international price. The import of gold into France is formally prohibited, but smuggling, particularly from Switzerland, is so simple that the price of gold in Paris differs little from the London and Zurich prices.

Gold smuggling has a multitude of ways and means. The stories about it would provide material both for a thriller or for a collection of anecdotes, although it is a serious business, and the syndicates even say that they are performing a socially useful job by uniting the demand for gold with its supply.

A stream of gold flows from Western Europe into the trading centres of the Middle East. At this stage it is not usually smuggling, since many countries in this region permit the import of the metal; from here it goes as contraband to neighbouring countries (Turkey, Iran and Egypt) or by the long, well-trod route to India and Pakistan. Its tributaries lead to Singapore, whence the gold wends its way as contraband to Indonesia and other neighbouring countries.

Another centre of smuggling in Southeast Asia is Hong Kong (Xianggang), a British possession on Chinese soil. From the tiny neighbouring Portuguese colony of Macao the gold is smuggled (almost openly) into Hong Kong, and makes its way from there to neighbouring countries. In the 1940s gold went to China: it was bought by officials, military men, industrialists and traders who were preparing to flee the country or hoping to survive the "hard times". At the beginning of the 1970s this was repeated in South Vietnam: in an effort to keep the valuables that they had plundered and stolen and to escape from the people's power, the ruling class and their lackeys bought up gold. Japan, the Philippines, Taiwan and South Korea have all been big clients of the Hong Kong smuggling syndicates at different times.

In the constant war against the customs authorities the smugglers and their bosses show miracles of organisation, cunning and inventiveness. The battle is usually decided in their favour. Of the many hundreds of tons of gold that cross the frontiers illegally each year only a trivial amount, less than one per cent, falls into the hands of the authorities. In 1947 the French authorities seized a ton of gold at the frontiers, while the annual turnover of the smuggling was estimated at 500 tons.[15] Japanese customs officials are more alert or more honest: in the particularly successful year of 1967 they seized more than 3 of the 30-40 tons of gold smuggled into the country that year. A large part of the gold that fell into the hands of the authorities was concealed in barrels of oil products with false bottoms that had been shipped to Japan from Canada.[16]

The history and present-day practice of gold smuggling has a multitude of devices. There are the usual, regularly used methods. Couriers flying on regular air flights wore under their suits special jackets with pockets designed to take 40 kilograms of gold on one's body; obviously only strong, young people were suitable for this. In recent years special checking of passengers has put an end to this method. Gold is usually smuggled into India by motor boats from ports in the Persian Gulf, and into Indonesia by junks from Singapore. Unusual, exotic and sometimes anecdotal methods and incidents are also known. Gold enters Turkey from the neighbouring countries in the stomachs of cattle driven over the border. One Middle Eastern country sent almost solid gold furniture to equip its embassy in India.[17]

Smuggling is just one sphere of investing capital. It has a somewhat higher risk, but the profits are high too. Consequently no laws or police measures are able to stop the smuggling of gold.

Hidden Gold: Hoarding

The 18th-century French philosopher Montesquieu included in his work *De l'esprit des lois* a small pamphlet against defenders of slavery. He mocks their "arguments" in favour of slavery among which is the following: "One proof that negroes have no common sense is that they prefer a necklace of glass to gold, which is so greatly treasured in the enlightened nations."[18] In the following centuries gold has continued to be treasured by the enlightened nations and has spread its power together with the advance of enlightenment to the less civilised nations. The hoarding of gold as treasure has been known since time immemorial, but here we shall speak of its forms, motives and consequences under capitalism in the latter half of the twentieth century.

As already mentioned, the total amount of gold hoarded in the capitalist countries is estimated at 26,000 tons, and this amount is being supplemented annually by an average of not less than 300 (and perhaps even 400) tons, the largest accumulations being in France and India. Let us consider the question more closely.

The word "hoard" means to amass money. Consequently paper money can be hoarded too. But with steady inflation not even the most dull-witted person would do this. The depreciation of paper money in recent years in the majority of countries is of such dimensions that one need only keep a banknote for five or seven years for it to lose half its purchasing power. And this is one of the main motives today behind the hoarding of *gold,* the price of which, as experience shows, can usually increase in paper money.

Hoarding is not all the same. To such an extent that in recent years, alongside the traditional, "classical" hoarding, so to say, of the Indo-French type, two other forms have been distinguished: the accumulation of gold by private owners as a form of *investment* of money capital and as a means of *speculation*. Naturally the dividing line between all three forms is blurred and indistinct, but at the same time the differences are considerable.

The traditional hoarder buys gold in order to *store* the value. He does not intend to part with it and, as a rule, does not do so unless this is absolutely necessary, even if it could be done to some profit.

The investor and the speculator buy gold in order to *increase* its value. For them it is investment of capital not different basically from other forms of investment: securities, land, and commodities such as grain or copper. They differ from one another only in that the investor invests the capital for a comparatively long period, and the speculator for a short one, intending to sell the gold when its price goes up. But this difference is not a fundamental one, and statistics, if they exist at all, usually combine these two forms. The investment of money capital in gold and speculation in gold did not develop seriously until after 1968, when the free market prices began to fluctuate sharply and soon soared up. Before this, as we know, the official price of gold (in dollars) had remained unchanged for several decades, and the free price hardly differed from it.

The investor and the speculator are often different social types from the traditional hoarder. In many cases they are big capitalists with access to the stock exchange, the real estate market and bank loans. It is quite possible that they are not even individuals, but juridical persons—firms and banks.

In the novel *The Moneychangers* by the modern American writer Arthur Hailey, the action takes place in the mid-seventies in the world of American banks and multinational corporations. One of the characters, the publisher of a financial bulletin for a narrow circle of rich people, Lewis D'Orsay, a conservative and adversary of new-fangled theories about the demonetisation of gold, advises his clients to invest up to 40 per cent of their capital in gold and gold-producing company shares (their price usually goes up with a rise in the price of gold).[19] From his observations of real life, Green writes that "there are plenty of multinational companies who like to keep a little reserve in gold, against a rainy day".[20] Among present-day hoarders one can find dictators from small Latin American countries, "oil sheikhs" who have recently grown rich, religious leaders and stock exchange sharks, film stars and bosses of organised crime. When the Greek shipowner millionaire Onassis had got some "spare money", he bought a million ounces (31 tons) of gold and sold it later at a modest profit.

In recent years banks, dealers and markets have worked out many ways by which rich people and firms can buy, hold and sell gold. Holding gold is risky and inconvenient, particularly when it is a question of large amounts. Today there is no need for this. The large banks in Western Europe, North America and other parts of the world carry on extensive operations to accept gold from clients for storage. In this case its sale amounts to conceding the right to draw it from the bank, to handing over a receipt for the gold stored. It may change hands many times without moving physically from the vault of the Zurich, London or New York bank. Deals involving the buying and selling of gold are carried out instantly with the help of the most advanced technology in communications and banking settlements. Big speculators buy and sell gold in different commercial centres and try to profit from the slightest difference in the price. Thanks to the difference in the time zones the organised gold markets (commodity exchanges) in the different centres (Western Europe, the USA and Canada, Singapore and Hong Kong) function at different times in the day and night. Speculators make use of this to deal all round the clock. As on the foreign exchange markets, there is *arbitrage* (constant purchases on some markets and sales on others) on the gold markets, thanks to which the prices on the free markets, where there are no restrictions and taxes, level out and form a more or less single world price. If gold is sold not for dollars, but for some other currency, the exchange rate in relation to the dollar plays a part in the price formation. If gold is bought with loaned resources, the rate of interest which must be paid on the credit is important. Dealings involve rapid complex calculations which today are often done by computers.

If gold is kept in a bank its owner receives a document certifying right of ownership, a gold certificate. This also exists if a client buys gold from a bank but does not remove it physically. In recent years this form of gold ownership has been widely used in the developed capitalist countries. There are two types of certificate: for allocated and non-allocated gold. In the first case, the client, as it were, keeps the gold in a safe, he can come to the bank and admire his property. He is allocated certain bars from the bank's gold stock. In the second case, the client has the right to the bank's gold in exactly the same way as a depositor has a right to his savings, i.e., to impersonal gold defined only quantitatively. The bank is able to put the gold deposited with it into circulation in the same way that it circulates the money capital deposited with it. All it needs to do is to keep a certain reserve in the event of gold being demanded by a certain number of clients. The holder of a certificate may sell it to another person, who thereby acquires a claim on the bank. Thus, a new credit document arises and develops, which always represents a fixed quantity of gold. Under inflation the long-forgotten classical banknote exchangeable for gold is being resurrected, as it were, in this peculiar form.

In this case the gold that is actually in the bank becomes the basis for a much larger circulation of gold certificates, because the bank can issue several times more certificates than it has gold deposits. The size of these operations rose particularly sharply after 1975 and is quite likely to rise even further. Today's investment advisers, real people, not fictional characters like the above-mentioned D'Orsay in Hailey's novel, as a rule advise holding a certain percentage of money capital in gold or gold certificates, and with the growth of inflation and the deteriorating position of the dollar this percentage is tending to increase.

The gold in the hands of big investors and speculators also serves as a basis for the development of forward transactions of a purely speculative nature. The essence of a forward transaction is that gold is sold with a future delivery date (from a few days to a few months) at a fixed price. This price is established in the form of a premium or discount on the spot price. For example, the price being 300 dollars an ounce the speculator buys gold to be delivered three months later at a premium of 10 dollars. If the price is 315 dollars in three months' time, he gains 5 dollars an ounce, because he can straightaway sell the commodity delivered to him at 310 dollars for this price. If the price is, say, 305 dollars, he loses 5 dollars an ounce. Actual delivery of the gold at the agreed time does not usually take place, and all that happens is that the loser pays the winner the difference. This is called playing the market.

One of the most important features of the gold market in recent years has been the tremendous development in futures trading. The main centres of this trade are in the United States: the turnovers of the gold sections in the commodity exchanges in New York and Chicago far exceed the scale of dealings in other countries. These operations did not begin in the USA until 1975, when the 1930s legislation prohibiting the private ownership of gold was repealed. A huge business grew up in a few years, the scale of which expanded greatly in 1978-1980. There are also exchanges where gold futures are contracted in Winnipeg, London, Zurich, Singapore and Hong Kong. But they are of secondary importance to the American exchanges.

Gold futures are primarily a sphere of speculation like futures with shares on stock exchanges and similar deals with certain other commodities. The sharp fluctuations in the price of gold in recent years, almost completely unknown until the seventies, have turned gold into a favourite object of speculation: at the height of the gold rush in the winter of 1979-1980 the price of gold fluctuated as much as 50-60 dollars a day.

The commodity exchanges where gold deals are transacted are actually joint stock companies run by a board elected by shareholders. The deals on the exchanges are transacted only by members of the exchange (predominantly large dealer firms). They transact the deals on their own behalf or by proxy. Futures are transacted for amounts of gold, multiples of a contract unit of standard size. On Chicago's International Monetary Market (division of the commodity exchange) the standard trading unit is 100 fine troy ounces. For example, a deal may be transacted to sell 20 contracts of 100 ounces each to be delivered in three months' time. The trading rules at the exchanges also regulate certain other conditions of transactions.

The scale of operations on forward markets is very great. In 1978 the turnover in futures on the New York market was 20,600 tons, which is twenty times more than the whole annual gold output of the capitalist countries.[21] Considering the high and rapidly rising prices, it must be estimated in terms of money at many thousands of millions of dollars. In the comparative calm of 1978 only 2 per cent of transactions ended with the actual delivery of the metal by the seller to the buyer, the other 98 per cent were settled by paying the difference. Thus, in one year about 400 tons changed hands as a result of these transactions. This transfer is also usually effected without the physical movement of the gold. It simply means that the name plates on the bars are changed in the special storehouses of the exchanges where dealers' reserves are kept. The total size of these reserves is considerable: on the New York and Chicago exchanges it was 90 tons at the end of 1978, and it has probably grown since then.[22]

The forward gold market played an important role in the gold rush of the winter of 1979-1980. This is what happened. The price of gold had been rising for several months in 1979 and in the autumn many big speculators began to speculate on it dropping: to sell gold with a delivery date in December or January at a discount, i.e., at prices lower than the current market prices. These speculators thought the US Treasury would sell a large amount of gold at its regular auctions and that this would result in a sharp drop in prices which, at the level of around

400 dollars an ounce, seemed excessively inflated. If this had happened the buyers would have had to pay the punters, as *The Economist* magazine calls them, the difference between the contract price and the actual market price at the moment when the deal was made. However, the magazine continues, these speculators "were caught with their breeches down", because the Treasury cancelled the auctions, and the price of gold went up instead of down.[23] In the universal pursuit of gold many buyers demanded the real gold delivery when the contract date expired. At the New York exchange in December 1979, 9,000 contracts ended with the actual delivery of the metal, which was three times the November figure. At the Chicago exchange the proportion of these contracts was 7 per cent instead of the usual 1-2 per cent. Speculators were forced to buy up gold on any market in order to honour contracts. This added fuel to the flames of the gold rush and sent prices even higher.

The traditional hoarder is usually a simpler case. He knows little about the exchange, does not read financial bulletins, and does not transact futures. The petty and middle urban bourgeoisie, the wealthy peasantry, certain strata of the intelligentsia and the workers' aristocracy—this is the social environment in which it is customary to hold a part of one's savings in gold. Although the name of these hoarders is legion, the vast majority keep only a few ounces of gold in small bars, coins or crude articles in cups, under mattresses or tucked away in other places. It has been estimated that in India 90 per cent of all hoarded gold is in the hands of 3 per cent of the population.[24] René Sédillot limits himself to this elegant, but rather meaningless statement: "Who hoards in France? The most modest and least educated savers, rather than experienced capitalists. Peasants rather than city folk, workers rather than specialists. On this matter, one should not pay much attention to opinion polls: the French are too fond of secrecy to reveal their true preferences to an interviewer."[25]

The vast majority of gold owners try to conceal the fact. In many countries such accumulations are illegal or not entirely legal. There is also, naturally, the fear of being robbed or blackmailed. But there is yet another important factor: gold is an ideal means of tax evasion. It is a good way of concealing incomes which for various reasons a person does not wish to divulge.

It is a better protection against the death duties which exist in most countries than investments in real estate or Old Masters.

There is much that is irrational, illogical in the behaviour of the small hoarder, in his attitude to gold. John Maynard Keynes, the famous English economist who wrote a great deal about money and gold, quotes the authority of Sigmund Freud, who found that a passion for gold lies in the human subconscious.

To our mind the sources of people's striving to accumulate money and gold cannot be deduced from the biological nature of man, from his physiology. Gold was not always and will not always be money in human society. Money itself is characteristic only of certain concrete forms of the structure of this society. The passion of money is *social* in nature. It is closely connected with the nature of a society in which money is transformed into capital and becomes the measure of all things, and of man himself. These social conditions turn gold into a fetish, and where there is a fetish, little room remains for rationality and logic. This position helps to explain the many paradoxes and eccentricities of hoarding.

In the Arab East, English gold coins, sovereigns, are particularly popular. Coins with kings' heads cost more than those with the heads of Queen Victoria and Queen Elizabeth. In France bars of gold weighing a kilogram are very popular with rich hoarders. They are affectionately called *savonettes* because they have the size and shape of a bar of toilet soap.

The not so wealthy hoarder prefers gold coins: they are more customary, more portable, and more attractive. With coins there is less risk of forgery and they are easier to sell should the need arise. All this has to be paid for, and gold costs more in coins than in bars (it is sold at a premium). The reserve of old, genuine coins that were struck during the period of the gold standard is limited, of course. Therefore, in order to profit from the higher price of coins, the state mints of many countries are issuing new coins, which are either a precise reproduction of the old ones, or correspond to them in gold content, but bear a different representation and year of issue. These coins are called official coins. In recent years, when the demand for gold and its price have risen sharply, the minting of official coins has increased. In three years (1974-1976) more than 700 tons of gold was used for this. Unlike previous years a very large number of the new coins were bought up not in France and India, but in the USA and Great Britain where hoarding is becoming fashionable.

First place in the world for the minting of gold coins belongs to the Republic of South Africa, which a few years ago introduced a new coin onto the world market—the Krugerrand. This word is formed from the name of the President of the Transvaal during the period of the Boer War, Stephanus Kruger, and the name of the present monetary unit of South Africa, the rand, which, in turn, is abbreviation of the Boer word "Witwatersrand" (the main gold-producing area). The Krugerrand contains exactly one troy ounce of pure gold, which makes it particularly convenient for buying and selling. This coin does not have a face value, as it were, only a weight. The price of an ounce of gold in Krugerrands is only a few per cent higher than its bar price, but in vast quantities this gives South Africa a fair profit.

In 1978-1979 the Krugerrand reigned almost supreme on the market of the United States. Of the 6.1 million coins sold in 1978 by South Africa more than half went to the USA. This means that in Krugerrands alone Americans bought up about 100 tons of gold. South Africa sold 27 per cent of its annual output (194 tons) in this form. The main reason for this high demand was the weakening of the dollar and the desire of money owners to find a reliable method of investing it. At the same time the extensive advertising campaign organised by firms selling Krugerrands is also of importance.

In spite of the large-scale production of Krugerrands they were very scarce in the period of the "great Christmas rush", the winter of 1979-1980. The South African Mint closed as usual for the Christmas and New Year holidays, after guaranteeing the usual supply of coins for sale. But demand exceeded all estimates and expectations, and the reserves were exhausted. As a result the premium on Krugerrands rose considerably compared with the bar price of gold.

Many other countries mint official coins, but none of them comes anywhere near South Africa in volume of production. In 1978 Mexico ranked second, spending about 23 tons of gold on minting coins. Mexican coins are very popular in France, and also sell well in the USA. In Britain a considerable number of sovereigns are minted, and orders are accepted from other countries for the minting of their national coins.

In autumn 1979 Canada started to mint coins similar in all respects to the Krugerrand, i.e., containing exactly one ounce of pure gold. Because of the representation of the national emblem of Canada on it, this coin became nicknamed the "maple leaf" on the market. There are reports that Australia also plans to mint and sell coins with a content of one ounce of pure gold. South Africa has put a coin on the market worth a quarter of Krugerrand, i.e., containing about 7.7 grammes of pure gold. It is designed for small hoarders who find it difficult to buy an ounce of gold at today's prices. The Swiss Credit Bank sells standard bars weighing only 5 grammes, the size of a postage stamp. In this way sellers of gold hope to reach the "man in the street", to persuade people with medium or even low incomes to buy gold.

The minting and selling of coins is a profitable business, and private capital, naturally, cannot ignore it. In many countries, particularly in the Middle East, coins that are copies of well-known and popular issues are minted, from mediaeval Turkish ones to new sovereigns. The gold content in these coins usually corresponds to the official standard. Incidentally, there have been cases of forged coins, usually with a lower content of pure gold.

Various types of memorial and jubilee medals are minted from gold. Until recently a relatively small amount of metal was used for this purpose. But in 1978 the United States Congress decided that for a five-year period an annual amount of one million ounces (31 tons) of gold should be allocated from the Treasury reserves for the production of gold medallions bearing portraits of Americans who are famous in the arts. It is assumed these medallions, the production of which will take more than 150 tons of gold, will be a rival to the Krugerrands.

Countries in which the Olympic Games are held traditionally strike special coins which quickly acquire collection status and are sought after by coin collectors. These coins are frequently made of gold. In connection with the Summer Games in Montreal in 1976 the Canadian Mint struck one million commemorative coins with a face value of 100 and 150 Canadian dollars from 10.5 tons of gold. These coins soon acquired a considerable premium against the price of bar gold.[26]

The Soviet Union also minted Olympic coins on the occasion of the Moscow Games in 1980. As well as copper-nickel, silver and platinum coins some commemorative gold coins with a denomination of 100 roubles were issued. They are official coins with a mandatory paying power in accordance with their face value. In other words, theoretically any payment of 100 roubles can be made with this coin. Naturally they too have become sought after by collectors.

Industry and Art

The first gold objects known to archaeologists date back to the fourth millennium B.C. (Egypt, Mesopotamia). A few years ago in Astrakhan Region some gold articles of the third millennium B.C. were found—to date the most ancient in Eastern Europe. The treasures of Graeco-Scythian art in the Leningrad Hermitage are world-famous. In 1969 the Museum of Historical Treasures of the Ukrainian Soviet Socialist Republic opened in the former Kiev Monastery of the Caves. It contains gold articles by Graeco-Scythian and early Slavonic masters, including some remarkable recent finds in the Ukraine. The world's museums treasure the work of ancient and mediaeval jewellers. In modern times too gold has been worked by highly gifted masters. At the same time there gradually developed the artisan and then the factory production of jewelry; in the 19th and 20th centuries the jewelry industry developed on a capitalist basis and began to work for a mass market.

Until the 20th century jewelry-making was essentially the only sphere of the industrial, non-monetary use of gold. It is estimated roughly that this sphere absorbed 22 per cent of all the gold produced over the period 1860-1913.[27]

The consumption of gold in the form of jewelry is a rather complex socio-economic and psychological phenomenon. Gold satisfies man's aesthetic needs, but since time immemorial it has also been a visual expression of social differences: the highborn and rich have gold ornaments, the lowborn and poor do not. In the Middle Ages, when the dividing lines between the estates were clear-cut and people's lives strictly regulated, there were often rules that prescribed how much gold a family belonging to a certain estate could possess and in what form. Until quite recently a person's wealth and social status were determined by his ownership of a bulbous gold watch, the thickness of its chain, and the gold and jewels of his tie-pin and cuff-links. Things have changed to some extent, of course. Today a millionaire, a manager of a big corporation or a minister may not wear a gramme of gold. New material indicators of social status have appeared. In the spirit of the times gold has been "democratised", wedding rings, bracelets and earrings of gold are produced by the millions.

But this did not change the essence of the matter. First, in certain social circles in the West gold is still an object of prestigious, ostentatious consumption. From time to time one hears of gold baths and door-knobs in the mansions of the rich, and of gold dog collars. Likewise an ordinary businessman may bring out a gold cigar case or cigarette lighter in the course of a business talk with some pleasure and to some effect. A gold watch is the standard present from a firm to an employee who has given twenty-five years of faithful service. Secondly, apart from the West, there are many Eastern and Southern countries where the bourgeois nouveaux riches, rich landowners, and prosperous officials and lawyers are only too glad to "consume" gold. At one time the Indian maharajahs were the best customers of the European jewellers. Now their place has been taken by the wealthy oil sheikhs.

Gold ornaments, however, have not always been displayed. On the contrary, they have often been hidden. As we have already mentioned, there is no clear distinction between jewelry and hoarded accumulations. To the family of an Indian peasant the gold bracelet of the wife or daughter is an insurance policy against a poor harvest or sickness. In India 600,000 jeweller craftsmen are employed in the making, buying and selling of gold and silver ornaments. Those who are wealthier engage in usury also, giving loans for pawned valuables. Here gold articles are essentially a form of hoarding. And the following is an example of the reverse: in Saudi Arabia, which absorbs a mass of gold coins, women's necklaces in the form of sovereigns threaded on a chain are in fashion; hoarded gold is thus turning into ornament.

The consumption of gold in the jewelry industry grew rapidly in the fifties and sixties under the influence of two main factors: first, the growth of the economy and incomes; secondly, the stability of the price of gold at the time of an overall rise in prices, which made it a relatively cheap form of raw material. The fivefold rise in the dollar price of gold in 1972-1974 reduced demand. The economic recession of 1974-1975 had the same effect.

As a result of these two factors the demand of the jewelry industry for gold dropped sharply. Whereas at the beginning of the seventies it consumed an estimated thousand tons, in 1974-1975 the average annual figure barely exceeded 350 tons. However, the jewelry industry managed to adjust to the new level of prices for its main type of raw material and successfully shifted sharp rises in costs onto the consumer. In 1976-1978 this was helped by an upswing in the economy of most industrial countries. But of even greater importance was the new rise in inflation, which made people hunt for reliable valuables. Not only in the East, but also in Western countries the investment and hoarding element in the buying of jewelry increased. As a result during this period the jewelry industry again began to consume up to a thousand tons of gold a year. This amount abruptly dropped in 1980 owing to the sharp rise in price.

The period 1977-1978 was marked by a long and significant fall of the dollar in relation to more stable currencies, particularly the West German mark, the Swiss franc and the Japanese yen. Thanks to this the price of gold in the above-mentioned currencies rose far less than in dollars, and sometimes even dropped. For jewelry firms in these countries gold did not grow more expensive, but in fact became cheaper insofar as the prices of other commodities continued to rise. We shall consider this question in more detail in Chapter IX. Here let us merely note that this factor played an important part in supporting the high demand for gold by the jewelry industry of these countries. The same also applies to other spheres of the industrial use of gold. In Japan industry consumed 75 tons in 1975, and more than 100 tons in 1978 (the jewelry industry accounting for 39 tons and 57 tons, respectively). In West Germany the industrial use of gold rose from 58 tons to 92 tons, and in Switzerland from 18 to 33 tons.[28]

First place for the amount of gold used in the jewelry industry, however, firmly belongs to Italy. Its share is usually about 20-25 per cent of the total for all capitalist countries. In Italy the advantages of large-scale production with a high degree of mechanisation and a modern system of management were introduced in the jewelry industry earlier and more effectively than in many other countries. The largest producer of jewelry in Italy and in the capitalist world as a whole is the firm of Gori and Zucchi, whose enterprises are in the town of Arezzo near Florence. At the beginning of the seventies this firm was using more gold than the jewelry industries of Great Britain, France and Switzerland put together. This firm has its own refining plant, which brings the gold up to established standards, and itself produces semi-finished goods used in the manufacture of jewelry. The firm has the most up-to-date equipment. It employs 30 designers who make it a fashion-setter in jewelry.[29]

Gori and Zucchi and other Italian firms produce a vast amount of jewelry designed for the mass consumer. Only a small part is consumed in Italy and most of it is exported. Italian gold articles are highly competitive. They have won markets in other West European countries and the USA, and also markets in the East.

About one-third of all the gold used in the jewelry industry is consumed by the countries of the Middle East, including Turkey. Alongside the traditional small-scale production of local craftsmen, modern enterprises built by Italian, West German and other Western firms have appeared in these countries in recent years. Taking advantage of cheap labour and relatively low taxes, they have in a number of cases ousted both local craftsmen and imports from the industrially developed countries.

In a great number of countries, including those of the West, hand-made jewelry has an age-old tradition, and this tradition has proved to be extremely long-lived. Even today a very large proportion of gold ornaments is made in small workshops mostly by hand. But as a rule the supply of semi-finished products (wire, sheet gold, etc.) to the small jeweller is now in the hands of big industrial firms (like the above-mentioned Engelhard Company in the USA). Jewelry workshops often take orders from big commercial firms. Thus small-scale production in the jewelry industry has ceased to be independent and has become part of the system of big capital.

The mass industrial production of jewelry has made the figure of the individual jewelry designer who creates unique works, as artists create pictures and sculptors statues, almost an anachronism. However, it cannot be said that this splendid art has died. Alongside the designers working for big firms there are workshops where talented goldsmiths still create splendid things. These goldsmiths and the economic aspect of their activity are described by Timothy Green in his book.[30] Unfortunately their work hardly ever reaches museums, but usually disappears into millionaires' private collections. Incidentally this applies not only to jewelry, but also to other works of art.

For the manufacture of jewelry gold of different fineness is used—from 8 carat (333 fineness) to 23 carat (958). In most countries there is state regulation of gold fineness for the jewelry industry and a system of checking and verification by state bodies of the fineness of articles produced. In Eastern countries, where the buying of jewelry has always contained a considerable element of hoarding, articles of high fineness are bought and sold (usually 21 or 22 carat). In many West European countries the most common fineness of gold articles is 14 or 18 carat. In Great Britain, West Germany and the United States many articles of low fineness, 8 or 10 carat, are sold.

The use of gold of different fineness and the use of various admixtures in alloys (silver, copper, nickel and palladium) is also connected with the fact that articles of a wide colour range are made of gold—from red (75 per cent gold and 25 per cent copper) to white (various percentages of silver or nickel). This increases the variety in jewelry on the market.

Archaeologists tell us that the first gold dentures were made for the nobility in Ancient Egypt and Sumer. The Romans knew of this use of gold. French physicians in the 18th century wrote manuals on this subject. Although in the 20th century stainless steel, plastic and other materials have taken the place of gold to some degree, it is still used in large quantities in dentistry and is considered attractive and prestigious. According to statistics for 1940, 76 per cent of American males and 56 per cent of American females had gold in their mouth in the form of fillings, crowns, bridges, etc.[31] The United States still holds first place for the use of gold in this sphere (usually more than 20 tons a year), but in per capita consumption Switzerland leads,

7-437

and in recent years West Germany and Japan have been moving up.

The total amount of gold used by dentists is slowly but steadily growing, and the creation of new materials is hardly influencing this process at all. In recent years 80-90 tons of gold has been spent on this annually, which is 8-9 per cent of the annual gold output in the capitalist countries. This is certainly a considerable amount.

Gold is also used in many other spheres of life, sometimes in strange and amusing ways. In the post-war years Japan, now the second industrial power of the capitalist world, has become a big consumer of gold. The management of a hotel just outside Tokyo installed a solid gold bath fashioned in the shape of a phoenix and made from 143 kilograms of gold. In the early seventies two minutes in this bath cost 1,000 yen (about 3 dollars).[32]

Table 4

Industrial Use of Gold in Capitalist Countries (in tons)

	Total*		Jewelry industry	
	1977	1978	1977	1978
All countries	1,405.1	1,552.2	995.6	1,000.8
Italy	220.0	249.4	209.0	235.0
South Africa	92.6	197.4	2.0	2.0
USA	169.0	178.3	79.5	83.4
Japan	75.0	100.2	48.5	56.7
West Germany	88.0	91.6	44.0	47.0
Turkey	91.0	91.5	80.0	86.0
Spain	54.0	57.8	49.7	51.5
Great Britain	45.4	49.3	24.0	22.0
Saudi Arabia and North Yemen	41.5	39.0	34.5	34.0
France	38.3	37.8	26.2	25.3
Mexico	13.4	37.5	7.9	9.3

* Including the minting of coins and medals.

Source: *Gold 1979*, Consolidated Gold Fields, London, 1979.

Since then inflation has probably increased the price several times over. This procedure is very popular, because many people believe in the curative effect of bathing in a gold bath. Belief in the curative properties of gold is very ancient and was cultivated by the alchemists of the Middle Ages. In the East village doctors and quacks still prescribe "gold pills" for many illnesses, including impotence and infertility.

Incidentally gold is used in modern medicine, but on a scientific basis. It is now used to combat cancer: in serious cases doctors prescribe injections of a colloidal suspension of radioactive gold.[33] The total amount of gold used in medicine is insignificant, however.

A most important sphere of the modern industrial use of gold is electronics, and here Japan plays an extremely important role. It is only slightly behind the United States in the amount of gold used in this sector. Considering that in the USA the produce of the electronics industry includes a high percentage of military equipment, it can be assumed that Japan is ahead of the USA in the use of gold in electronics technology for civilian purposes. These two countries taken together account for two-thirds of all the gold used in the electronics industry of the capitalist countries.

The unique combination of gold's physical properties—electroconductivity, anti-corrosiveness and plasticity—has been extremely important for electronics. Gold is used where a stable, durable and dense material is required: in electronic computers, space apparatus, and underwater communications. Sédillot remarks wittily: "The future belongs to machines with gold brains."[34] This reminds us yet again that in everyday speech the word "gold" has for centuries meant "excellent", "first class".

The following statistics relate to the USA for 1974: of the 44 tons of gold used in electronics, 21 went on contacts, 5 on printed circuits, 7 on semi-conductors and 4 each on cathode-ray tubes and switches. In the following two years the consumption of gold by this branch dropped to less than half this amount: this was explained by the introduction of substitutes which were developed as gold became more expensive, and also by the decrease in economic activity and fall in demand. But the experts forecast that the reserves of gold economy in electronics have been largely exhausted and very often the further introduction of substitutes will be possible only at the cost of inferior quality. Moreover, many countries have still to pass through the stages of development in the electronics industry that the United States and Japan have covered. Enterprises are managing somehow or other to adjust to the higher level of prices, just as the economy of the capitalist countries is adjusting to frequent rises in the level of oil prices. For all these reasons it is considered that the electronics industry will steadily consume about 100 tons of gold a year, perhaps even more.

There are many other spheres of the industrial and artistic use of gold. Here we find both traditional activities like the gilding of church domes and spires or gold embroidery. And more recent, but technically simple uses such as fountain pen nibs and spectacle frames. Then there are also entirely new spheres and methods of use, like electronics, connected with the modern scientific and technological revolution. Apart from its anti-corrosiveness and plasticity, gold possesses yet another important property: a high reflective capacity in relation to heat and light radiation. There is a thin layer of gold on the exhaust pipes of the Rolls Royce engines which are fitted in the supersonic Concorde: this prevents them from overheating. For the thermal protection the American Skylab station has a special screen made from aluminised and gilded plastic on the side facing the sun.

In all about 75 tons of gold was used in 1978 for industrial and decorative uses excluding electronics in the capitalist countries, about half of this in the USA. This demand is more stable than the demand of jewelers and electronics. It varies little from year to year, in spite of changes in prices and business activity. In economics this is called *low elasticity of demand*. Specialists think it will rise roughly in pace with the increase in industrial production.

Money and Non-Money

In concluding this chapter I should like to discuss two important questions. The first is the prospects for the use of gold and the demand for it on the world market. The second is the relationship of the demand for gold as money and gold as a commodity (non-money).

Over the last century people writing about gold often maintained that if gold ceased to be money the demand for it would drop, which would cause a sharp fall in its price in paper money and its exchange value in relation to other commodities. One such prophecy (by Paul Einzig) has been mentioned above. If today gold has not actually ceased to be money (a debatable point), at least in the last fifteen years the monetary sphere has shown very little demand for it. Its price has risen several times, however, and stands now at a record high level, although it sometimes fluctuates rather sharply.

At recent price levels *the sphere of industrial and artistic consumption is capable of consuming current world output of the metal.* This is the conclusion of all experts who have studied the problem of gold in recent years. Current output is not enough for hoarding today. This means that in order to satisfy the demand of hoarders it will be necessary to bring gold out of the central monetary reserves. This is being done, but in very limited amounts. Another solution would be to increase output, but all specialists agree that there are no prospects of this in the capitalist world. In this situation the "burden" of balancing supply and demand is placed upon the price.

Gold is money. Gold is non-money. These statements seem to contradict each other. But this is not the case. In real life phenomena are not observed in their pure form, opposites come together and everything exists in development and change only. The dialectical method of cognition demands this view of reality. Therefore both extreme views are wrong: namely, that gold is still money to the same extent that it was during the gold standard, and that gold is no longer money at all.

Before it became money it was an ordinary commodity, non-money. After it became money, gold retained the function of a commodity with certain useful properties. Marx characterised this proposition as follows: "The use-value of the money-commodity becomes two-fold. In addition to its special use-value as a commodity (gold, for instance, serving to stop teeth, to form the raw material of articles of luxury, &c.), it acquires a formal use-value, originating in its specific social function."[35]

As we can see, these two use values, two forms of utility, two functions of gold (the commodity and the money) do not exist separately, but in close unity, they are constantly intertwining and interacting. In particular, the very utility of articles of luxury is to a large extent determined by the fact that they are made from a metal which is a money commodity.

The use value (utility) of any commodity is varied and does not remain unchanged in the course of historical development. At one time oil was used only to burn in the primitive brazier of the nomad grazing his herds in places where it came naturally to the surface. At the end of the 19th century it was already being transported over large distances and processed into kerosene and other forms of fuel and lubricating oils. Today oil is the prime raw material for thousands of very useful commodities.

Nor does the use value of gold, its social utility remain unchanged. The word *utility* is understood here, of course, in the economic, not in the ethical or physiological sense. From this point of view narcotics have a "utility", not to mention tobacco. Like narcotics, gold can be the cause of a person's moral degradation and ruin. However, if an object satisfies a need, real or imaginary, natural or unnatural, it has *economic utility*.

It is obvious that in the combination of the two use values of gold there is a shift in favour of the non-money one. But this is not all. The content and forms of manifestation of each of them have changed. The money function of gold is concentrated primarily in the sphere of the world market, international financial relations, but there too it has changed, as will be shown in more detail below.

Hoards once served as a reserve of the money supply, the monetary use of gold. Today it is rather opposed to the monetary use: gold is not leaving the sphere of hoarding, not returning to the monetary sphere. Gold, one might say, has acquired a new specific utility in these accumulations—as a special hoarding commodity, a means of insurance against inflation and socio-political upheavals. To what extent hoards retain their monetary character is a debatable and difficult question. Much depends on how we define money, which aspect of these accumulations we stress. From a certain point of view the accumulation of gold resembles the buying of gems, Old Masters or rare books. We can hardly call these objects money in any real sense. But as an object of accumulation gold is also similar to securities or a bank account, and this is close to money.

Let us note the following, however. As non-money, gold is

perceptibly different from all the above-mentioned objects in that it is divisible, of uniform quality, and highly liquid (that is, easy to sell) and, most importantly, in the long tradition that has made it a favourite medium for the accumulation and preservation of value, wealth. In other words, as non-money it is still closer to money. Yet, on the other hand, as money it also differs from the above-mentioned paper-credit valuables in its lustrous materiality which modern man is no longer accustomed to associate with money.

These arguments may appear complicated. But reality is a complex thing. By simplifying it to the principle of "yes" and "no", we risk losing the very essence of it.

V. GOLD GENOCIDE

The image of the *age of gold* conjures up a picture of splendid, happy times. A wise Frenchman once said, almost two centuries ago, that the age of gold was the age when gold did not reign.

But in the real history of class society—and written history knows only of such a society—gold and murder go side by side. The murder of individuals and mass destruction, the extinction of whole peoples—what in recent times has become known as *genocide*.

To seize gold that has been mined or amassed. To force people to mine gold, working them to death. These are the two main forms of gold genocide known to history. Of course, in the civilised 19th and 20th centuries its extreme forms are not encountered so often. But in this period too gold still brings people death, national and racial oppression, and the destruction of cultural values.

Why gold you may ask. Conquerors have always seized the land, cattle and other possessions of the vanquished. Cruel slave-owners forced people to toil on sugar, tobacco and coffee plantations. But gold is not cattle, or sugar, or tobacco. It is wealth in the most absolute and direct form. In it, in money, lie all other commodities, all possibilities, all power over things and people. In Pushkin's aphoristic verse we read:

> *"All is mine," quoth gold,*
> *"All is mine," quoth the sword,*
> *"I will buy all," quoth gold.*
> *"I will take all," quoth the sword.*[1]

The natural compromise in this argument is to *take* gold by the force of the sword, in order to then *buy* all. And this is what conquerors strove to do.

Not all of them were content with land, cattle and even slaves. Gaining an income from these resources demanded time, supervision and effort. As war booty gold is far more convenient. After taking easily portable riches in the form of precious metals, gems, etc., the conquerors of ancient and mediaeval times would give all else over to fire and sword.

There have been cases when a man starving to death has killed for a piece of bread. It is not so easy to imagine a murder committed by a well-fed, well-dressed person for a sack or even a hundred sacks of sugar. But at all times killing for gold was regarded as something almost natural.

Antiquity

The great gold country in antiquity was Egypt. The gold treasures of the pharaohs and priests amassed by the bitter toil of generations of slaves and by bloody military campaigns into the heart of Africa were a strong enticement to all conquerors in the third, second and first millennia B.C. The great warrior of the 24th century B.C. Sargon, King of Akkad (a state in Mesopotamia), in his pursuit of gold and copper conquered Mesopotamia and the Eastern Mediterranean, but could not vanquish Egypt. The amassing of gold continued until the Assyrian King Esarhaddon defeated and looted weakened Egypt in the 7th century B.C.

Esarhaddon's treasures, stored in the Assyrian capital of Nineveh, were seized less than a century later by the Babylonians. But they too did not possess these treasures for long. In the second half of the 6th century B.C. Babylon was captured and looted by the King Cyrus of Persia. His son Cambyses was not content with this and, after looting the pharaohs' treasures once more, he decided to conquer the lands rich in gold to the south of Egypt. There he perished together with his army.

Herodotus tells the following story about the Persians in Babylon. In the most lively part of the town was the tomb of a Babylonian Queen. Carved on the stone was the inscription: "If there be one among my successors on the throne of Babylon who is in want of treasure, let him open my tomb, and take as much as he chooses—not, however, unless he be truly in want, for it will not be for his good." King Darius of Persia, who also became ruler of Babylon, was tempted by this invitation. But when the tomb was opened instead of treasure he found only a stone with the inscription: "Hadst thou not been insatiate of pelf, and careless how thou gottest it, thou wouldst not have broken open the sepulchres of the dead."[2]

In the 4th century B.C. "all the gold in the world" was amassed in the treasure stores of the rulers of Persia. Already in antiquity legends arose about the fantastic amount of gold and silver that Alexander the Great of Macedonia captured after defeating Persia. Diodorus Siculus relates that in the three Persian capitals (Persepolis, Susa and Ecbatana) 340,000 talents of gold and silver were captured, which is about 10,000 tons. After the fall of Alexander's empire in the course of numerous wars and conquests this colossal treasure spread all over the civilised world of that day. One of the Hellenistic rulers of Bactria (a state in Central Asia) minted the largest known gold coin—a 20-stater weighing 168 grammes with a diameter of 58 millimetres.[3]

A new gold stock was set up in Rome, particularly after the conquest of Spain (2nd century B.C.), Gaul and Egypt (1st century B.C.). Julius Caesar was not as lucky as Alexander, but he too sent gold to Rome after all his campaigns. So much gold was brought from Gaul that it even dropped a quarter of its price against silver. Suetonius says that in Gaul he "looted the sanctuaries and temples of the gods, full of offerings; he destroyed towns, more often to get booty than to punish them for a fault".[4]

History shows that mines of precious metals have always been places of forced labour, where human life was worth nothing. Like the building of the pyramids, large-scale mining of lode gold in Egypt was possible only on the basis of slave labour. The Romans developed gold mining in Spain only after they sent thousands of slaves there.

Domestic, agricultural and artisan slavery never knew such terrible exploitation. But wherever *money* is being directly produced, greed and cruelty appear in their absolute forms, so to say, void of all that is humane. Marx said: "In antiquity over-work becomes horrible only when the object is to obtain exchange-value in its specific independent money-form; in the production of gold and silver. Compulsory working to death is here the recognised form of over-work. Only read Diodorus Siculus."[5]

Diodorus, a Greek historian of the 1st century B.C. and a skilled narrator, has preserved some remarkably detailed descriptions of Egyptian mines and the technological process of gold production during the Hellenistic dynasty of the Ptolemies. He describes crowds of almost naked people, men, women and children, working under the blows of the overseer's whip and having nothing to look forward to but an early death from exhaustion and illness.

The back-breaking work at the gold mines was the usual form of punishment for criminals in Hellenistic and Roman times. Probably far more early Christians perished there than were mauled to death by wild animals in the arenas, although writers and artists more readily turn to the latter because of its dramatic appeal. Later Christians treated pagans just as cruelly in the mining of gold.

Gold-mining methods in Spain are described by Pliny the Elder. His account is less emotional than Diodorus', but essentially paints an equally terrible picture. Pliny writes, for example, that people working underground did not see the light for many months. He speaks of the mortal danger of this work and the deaths of multitudes of people. Describing how the miners chiselled the hard rock with primitive tools, he remarks that only the lust for gold is more stubborn than this rock. Naturally he is referring to the lust not of the slaves, but of their owners.[6]

The White Man's God

The Cacique (something between a tribal chief and a feudal prince) Hatuey, who fled from the Spaniards when they captured the island of Española (Haiti) to Cuba, told the local Indians that the white man's god was gold. He urged the natives to throw all their gold into the river so that the white men would not find their god and would leave them in peace.[7]

Thus writes Bartolomé de las Casas, a Spanish priest, one of the early humanists and enlighteners of the Indians and the author of some remarkable notes on the first few decades of the conquest when millions of people were killed for the sake of gold.

It is difficult for us to imagine today, but gold was perhaps the main motive behind Columbus' discovery and opening up of the new lands, which were subsequently called America due to a misunderstanding. Gold is the leitmotif of Columbus' first letter from Española which he had just discovered. He wrote excitedly that the island's rivers contained gold and that there were evidently gold deposits in the hills too. The natives had no idea of the value of gold and would give it away for any trifle. Columbus wrote that he had founded the first Spanish settlement in "a remarkably favourable spot, and in every way convenient for the purposes of gain and commerce".[8]

After ten years of sailing and discovery, in a letter of 1503 to King Ferdinand and Queen Isabella from Jamaica the great seafarer described the riches of the recently discovered lands (it later transpired that this was a continent—the Central American isthmus), in the most glowing colours, and reflected: "Gold is the most precious of all commodities; gold constitutes treasure, and he who possesses it has all he needs in this world, as also the means of rescuing souls from purgatory, and restoring them to the enjoyment of paradise."[9] As we can see, this is not so far from the naive ideas of the Cacique Hatuey.

It turned out, however, that Haiti and the other Antilles and lands on the isthmus did not possess large deposits of gold, and this was one of the reasons for Columbus' rapid fall into disfavour. Soetbeer estimates that over the period 1493-1520 only about 22 tons of gold was mined in the West Indies, mainly on Haiti. For the sake of this gold almost all the native population of Haiti was destroyed. According to various estimates the island's population was originally between one to three million, and by 1514, Las Casas tells us, only 13,000-14,000 remained. If we bear in mind that during this period many people in other parts of Central America also perished in close connection with the Spaniards' pursuit of gold, we shall not go far wrong in saying that these 22 tons of gold cost 2 million human lives, that is, 100,000 lives per ton, or three lives per ounce. This is probably some kind of sinister record in the bloody history of the yellow metal. It is hard to read without shuddering Las Casas' account (confirmed by many eyewitnesses, incidentally) of the monstrous mechanism of extermination, the barbaric cruelties committed by the Spanish conquistadors: the whole-sale murders and torture, the slavery, the back-breaking work in the mines, and the starving to death of thousands of people.

Of his fellow-countrymen Las Casas wrote that they advanced with the Cross in their hands and with an insatiable thirst for gold in their hearts.[10]

The white man's god also destroyed the highly developed and interesting civilisations of old America, the Aztec state in what is now Mexico and the Inca state in Peru. The course of events was the same everywhere: first the Spaniards would seize all the gold they could lay their hands on, then they would force the enslaved Indians to toil in the gold and silver fields and mines. For the Indians of Mexico and Peru gold had hardly performed the function of money at all and was valued only as a metal for ritual and secular ornament. Like Columbus, the conqueror of Mexico, Cortez, and the conqueror of Peru, Pizarro, consciously and unconsciously exaggerated the riches of these lands. They had to do so in order to recruit men, to gain the monarch's favour and to get money from bankers and merchants to organise their expeditions. In fact the amount of gold articles was not so large in either of the states, and the annual output from the river sands was quite insignificant.

During the siege and storm of the Aztec capital, Tenochtitlan, the chronicles tell us, 240,000 of the town's population of 300,000 perished and the captors seized only 600 kilograms of gold.[11] Soetbeer estimates that in the following twenty-four years no more than 5 tons of the metal was produced in Mexico.[12]

The thirst for gold that made the Spaniards perform remarkable feats and commit monstrous cruelties amazed the natives. One of the Indian legends about the fall of Tenochtitlan says of the Spaniards that they search only for gold, disregarding other treasures even such as jade and turquoise. And also: "They grabbed gold like monkeys, shaking with pleasure, as if it had transformed them and illumined their hearts with a bright light. For in truth they do strive for it with an indescribable thirst. Their bellies are swollen with it, they hunger after it as if they were starving. Like hungry pigs, they lust for gold."[13]

All this was repeated with new revolting details during the conquest of the Inca state by Pizarro and his gang. The ill-famed ransom which Pizarro obtained for their ruler Ataualapa (who was executed nevertheless) was 5.5 tons of gold and 11.8 tons of silver. In the taking of the Inca capital of Cuzco the Spaniards seized 1.1 tons of gold and 15 tons of silver. The Inca state was devastated and their culture perished.[14]

At the same time as the first conquistadors were spreading the religion of the Cross and gold in America by fire and sword, in Spain itself there raged genocide of Moors and Jews, in which gold played its sinister role. In banishing the "infidels" from Spain the authorities prohibited them from taking precious metals with them. Some tried to take gold with them in their own stomachs and intestines, by swallowing coins pounded into powder. Having learnt of this, the Moors of Fez (now Morocco) where the Spanish Jews fled, would sometimes kill people and rip open their stomachs in the hope of finding gold there.[15]

Stories of gold being swallowed recur throughout the history and folklore of different ages and peoples in both tragic and comic form, up to Chekhov's story about the rich man who ate his gold with honey just before his death to annoy his heirs.

Pirates

In 1512 the Spanish ships which brought gold and silver across the ocean from America were attacked for the first time by pirates, French privateers. After this for almost three centuries on the high seas of the Atlantic hostile military vessels and pirates hunted ships with valuable cargoes, and Spanish and Portuguese warships hunted pirates. If the gold did fall into the hands of sea robbers, it would change hands many times, rarely without bloodshed. How many souls perished in these battles and skirmishes about which a multitude of good and bad novels have been written, it is impossible to say.

The Spanish viceroys kept the departure dates of vessels carrying gold a great secret, but English, French and Dutch spies did their utmost to find them out. After receiving some secret information, Queen Elizabeth I of England sent Captain Francis Drake around Americas to the Pacific Ocean to attack the Spaniards at a point where they felt quite secure: on the route from Peru to Panama. Drake carried out the mission successfully, seizing several dozen tons of gold, silver and precious stones from the Spaniards, and returned to England triumphantly in 1580 from the east, being the second to sail round the world. Philip II's unsuccessful naval campaign against England in 1588 (the defeat of the Great Armada) had as one of its main aims to nip in the bud English privateering and piracy on the gold and silver routes.

One of the most dramatic and bloody episodes in the struggle for gold took place in 1702 during the War of the Spanish Succession. An Anglo-Dutch squadron surprised a Spanish-French gold fleet hiding in the Bay of Vigo, on the northwest coast of Spain. The cargo was valued at 60 million pounds sterling, a huge sum in those days. In the battle about 3,000 men perished and almost the whole cargo, for the Spaniards blew up the ships rather than let them be captured by the enemy. Unfortunately they sank to the sea bed at a relatively great depth, so that the Spaniards' attempts to salvage their treasure were unsuccessful. In the 19th century joint-stock companies were set up to seek for the "Vigo Galleons" and books were written about their fate. Attempts to salvage them have been made in the 20th century too. But so far only a small part of the treasure appears to have been recovered.[16]

Robert Louis Stevenson's famous novel *Treasure Island* has, as we know, a happy end in which Jim and his friends find the trove of the pirate chief, Captain Flint: "And in a far corner [of the cave], only duskily flickered over by the blaze, I beheld great heaps of coin and quadrilaterals built of bars of gold... How many it had cost in the amassing, what blood and sorrow." The writer, who was possibly interested in numismatics, goes on to describe what coins there were: guineas and Louis d'ors, doubloons and double guineas, moidores and sequins.[17]

A word about mediaeval gold coins in Western Europe. The year 1252 is usually accepted as the birth of the gold coin of the modern age, when the minting of the *florin,* which contained about 3.5 grammes of gold, began in Florence. It was named after the town. In the 13th century the minting of coins with the same amount of gold began in Genoa and Venice. Venetian coins, which were called *sequins* or *ducats,* became very well known. Many states minted coins with these names. In Germany and the Netherlands a coin with the same or a similar gold content was called the *gulden*. These coins accompany the economic development of Western Europe right up to the 17th and 18th centuries. Gold coin issued mainly for international circulation was far less subject to spoil and devaluation than silver coin, and sometimes kept its gold content for centuries.

Other coins mentioned by Stevenson were minted mainly in the 17th and 18th centuries. The *Louis d'or* (i.e., "gold Louis") appeared in 1640 under Louis XIII and was constantly changing its gold content and standard up to the revolution at the end of the 18th century (under the three following Louis). The *guinea* (from the name of the country—Guinea) was modelled on the Louis d'or and first minted in 1661. The gold content of the Louis d'or and the guinea fluctuated between 6.5 and 8 grammes.

The *doubloon,* also known as the "double escudo" and "pistole", was the main Spanish gold coin from the 16th to 18th centuries and contained about 6.2 grammes of gold.[18] This coin also circulated in France and is probably known to many from *The Three Musketeers*. For example, D'Artagnan is sent to London to save the Queen's honour with three hundred pistoles in his bag. This bag would have weighed about 2 kilograms and be worth up to 150,000-200,000 francs in present-day currency (at the market price of weight gold).

The Gold Rushes

This term, which arose in the last century, has been given another meaning in recent times. When at the beginning of 1968 speculators, expecting a rise in the dollar price of gold, began to buy it up in large quantities on the London and other markets, this was called a gold rush. It has one feature in common with the gold rushes of the 19th century—the pursuit of chance, fantastic profit. But whereas at worst the speculators did risk their money (or rather risk losing the interest on their money, because a drop in the price of gold was obviously out of the question), the gold rushes of the 19th century cost the lives of many, possibly hundreds of thousands of people.

The discovery of gold in California, Australia and Alaska and the savage forms of settling these territories led to the extinction and dying out of their indigenous inhabitants. The most repulsive features of "free enterprise" were manifested fully in each of these epics: the freedom to exploit and cheat, the unlimited power of money, egoism and cruelty, racism, violence and crime.

In 1849-1850 tens of thousands of people from the eastern parts of the USA and Europe flocked to California. Death diminished their numbers on the journey: it is estimated that no less than 5,000 lost their lives in the summer of 1849 on the overland trek alone through what was then wild, unsettled country. Dysentery and scurvy were rampant in California itself too. And firearms were just as essential a piece of equipment for the gold prospector as his pickaxe and pan. With a population of a few hundred thousand in California 4,200 murders were recorded for the period from 1849 to 1854. Lynching, the killing of people by a crowd without a trial or investigation, was common practice. The sick, weak or infirm, including old people and children, had no chance of help. One eyewitness account reads: "You have heard of the Battle of Life—it is a reality here; the fallen are trampled into the mud and left to the tender mercies of the earth and sky... Money, money, is the all-absorbing object."[19]

Buyers-up frequently paid prospectors two or three times less than the official price which corresponded to the gold content of the dollar. Prospectors were fleeced by tradesmen and owners of the various establishments for entertainment. The library of the University of California in Berkeley has a copy of an advertising booklet published in 1849, inviting people to find easy gold in California; the author hid behind the initials D.L., but the hand of some unfortunate immigrant has deciphered these letters as "Damn Liar".[20]

Things were particularly difficult and the number of victims particularly high among the non-white participants in the gold rush—Indians, Mexicans and Chinese. They were deprived of the right to make their own "claims" to gold-bearing plots of land, and they were either hired by whites or worked on dumps containing remains of gold. Many cases of outbursts of racial violence were recorded. The forms of local self-government which arose among the white Anglo-Saxon population were aimed against the poor and people of other races and nationalities. One English writer wonders in this connection: "Was this, as nineteenth-century writers frequently maintained, a supreme manifestation of democracy, or was it, as twentieth-century critics might hold, an embryonic form of fascism?"[21] The gold discovered in 1851 in New South Wales and Victoria transformed the face of Australia, which had previously been nothing but a place of hard labour for English convicts and a vast pastureland. Thanks to the surveillance of the authorities there was perhaps less shooting and more order here than in California, but on the whole history repeated itself.

The first organised protest of Australian workers is connected with gold. Although the immediate cause was displeasure at the high taxes which the government levied on each "claim", the workers also advanced political demands similar to those of the Chartists in England: extension of suffrage, democratisation of Parliament, etc. In autumn 1854 in the gold-bearing district of Ballarat (not far from Melbourne) there was even an armed clash of workers with the police in which several dozen people were killed. This massacre, known as the Eureka after the hotel where it began, was for many decades the slogan of the Australian working-class movement, a warning to the ruling classes of the militant resolve of the unfortunate.

The last "classic" gold rush was at the end of the last century, when gold was discovered in North Canada and Alaska. In the 1898 season about a hundred thousand people set out from the coast to the gold-mining areas along the River Klondike. The road across the Chilcoot Pass and down the Yukon to the gold "capital" of Dawson City was long, hard and dangerous. It is estimated that not more than thirty or forty thousand reached their goal. As always, only a few hundred became rich.

Unlike California and Australia, gold did not bring life to these remote, Arctic lands. The exhaustion of the placers doomed them to oblivion. Dawson (the centre of the Yukon Territory of Canada) now has a population of 750; at the height of the gold rush it was 25,000. The town lives on the tourist trade: the notorious saloons have been restored, and there are various festivals and "amateur gold mining" with hiring of equipment used by old prospectors.[22]

From time to time there are "mini-rushes" caused by new finds of gold too, but they are of no economic importance. Known geological deposits of the metal are firmly controlled by the concerns.

Fascism and Gold

The policy of Hitler Germany in relation to gold was an integral part of the genocide, concealed with monstrous hypocrisy, which the Nazis carried out in relation to the peoples of Europe.

Germany spent practically all its gold reserve on preparations for war. At least, officially it possessed hardly any gold on the eve of the war; there was probably a certain amount of gold in secret reserves. In 1938-1939 the European countries hurriedly moved their gold to the USA, partly selling it for dollars, partly depositing it in the vaults of the Federal Reserve Bank of New York described above. F. I. Mikhalevsky, a Soviet specialist who has devoted many decades of study to the economic problems of gold, comments wittily: "Observing the flow of gold from Europe, the fascists must have felt like robbers who have pinpointed a flat for their next robbery and see all the valuable things being carried out of it."[23]

Nazi propaganda and official economics was against gold and against it retaining important economic functions. Gold and gold currency was represented as the product and bulwark of rotten "liberal capitalism". In the "new economic order" which the Nazis were trying to set up in Europe in order to extend it to the rest of the world there was no place for gold: everything would be paid for in Reichsmarks and everybody would grant free credit to Germany.

But propaganda and theory are one thing, and politics and real life quite another. In fact the Nazi leaders tried to grab every grain of gold that came within their sphere of influence. In each of the countries they conquered they sought to seize the gold reserves of the central bank if it had not been totally exported beforehand. They confiscated gold belonging to the population and looted churches and museums. At the Nuremberg Trial of the major German war criminals and at other trials it was reported that while killing millions of people in concentration camps these monsters carefully removed the gold articles belonging to them, including dental fixtures.

Keeping gold from the Nazis was a patriotic feat requiring great courage. The world press published the story of the gold reserve of the Bank of Norway, which was hidden by ordinary citizens and secretly taken out of the occupied country under the Nazis' noses.

The fate of the gold stolen by Nazis is largely unknown. Obviously only a small part of it was discovered when Germany was defeated. Probably a certain amount of gold was secretly removed to neutral countries or hidden in Germany. Many people's imagination is stirred by real or imaginary troves hidden by the Nazis in remote spots. The author of a special work on sunk treasures writes that valuables worth 160 million gold marks were hidden underwater by the Corsican coast, near the town of Bastia. There is even more definite information about troves hidden in an Alpine lake in Western Austria. The Corsican trove is coveted by the Italian mafia, and several mysterious murders have already taken place. It is rumoured that the Alpine troves are protected by a secret Nazi organisation, which also does not stop at murder. Gold continues to kill, even when it has disappeared out of sight, even just by its name.[24]

Apartheid

The word "apartheid" in the language of the Boers, descendants of settlers from Holland in South Africa, means division, separate existence. Ever since the 1950s apartheid has been the official policy of the racist regime in the Republic of South Africa. But this system developed far earlier.

In the West it is generally thought that apartheid was the creation of Boer nationalists, and that settlers of English origin had nothing to do with it. In fact it is the common position of white reactionaries and racists in relation to all non-whites, particularly to Africans. The gold-mining industry grew up as a predominantly "English" sector; only recently did Boer capital begin to play a part in it. But it is precisely in this branch that many of the revolting features of racial segregation and discrimination, i.e., apartheid, began to develop in the late 19th century.

The organisation of labour at the South African gold mines (and also in the uranium and coal mines) has no analogy in the modern world. The companies not only do not seek to form permanent, skilled personnel of African workers, but do their utmost to oppose this. They have created a system of short-term (not more than two years) recruitment of cheap and unskilled labour manpower from the remote areas of the country and neighbouring African territories, now independent states. The recruits are brought to Johannesburg and assigned to mines. As a rule they are illiterate and do not speak English. They do not know in advance where they are going to work, what their work will be, or even how much they will be paid. The "training" lasts three or four days, after which they are sent underground, often at a depth of two or three kilometres. The African carries out only the rough physical work. He has no chance of promotion or of improved working conditions. Trade unions and any other organisations of African workers are prohibited.

The African miners live at the mines in so-called compounds, a type of barracks, in rooms with two-tiered bunks for ten to twenty people. Their freedom of movement is strictly limited. They cannot have a family with them. Like the whole system of apartheid, the regulation of the work and life of the African miners bears the imprint of cold-blooded civilised barbarity. Humiliation of human dignity is here the norm, the established type of inter-racial relations.

The gold mines kill people. The South African scholar Francis Wilson estimates that over the period 1936-1966 no less than 19,000 men, 93 per cent of them black, died as a result of accidents in the gold mines.[25] The annual "norm" for the following period was 500-600 deaths. The death rate for Africans is about double that for white miners. The reason for this is well known: the blacks work in the most dangerous sections and are not so well trained. For each case involving loss of life there are on average about forty cases of permanent disability and severe injury. It is known that, apart from accidents involving loss of life or permanent disability, there are also thousands in which African miners contract such pulmonary diseases as pneumoconiosis and silicosis during their work. To this it must be added that they often fall victim to tuberculosis. Statistics are more eloquent than words: in 1972 the black miner on average obtained 24 rands a month for his dangerous and back-breaking work, whereas the average wage of the white worker was 391 rands, that is, sixteen times more. In no other industry was the gap so great. This does not exist, of course, in any other country in the world. The companies maintain that in addition they also spend more than an average of 20 rands a month on accommodation, food and medical care for each worker, but this does not change the essence of the matter.

Basically this is the concept of slave labour: just enough is spent so that a man can live and work for a year or eighteen months, as long as he is needed by the company. Payment in cash is obviously not enough to support a family: the family has to survive on its own meagre plot somewhere hundreds of kilometres from the mines. The head of a manpower recruitment organisation literally told Timothy Green the following: "He [the African] comes to get pin money. We've never said that our wage is sufficient for him and his family to live on, but he comes here to get money for extra luxuries."[26] It is a well-known fact that in practice these "luxuries" are tobacco and alcohol.

An important role is played in Marxist political economy by the concept of the value of labour power. It is determined by the costs needed for a worker to live "like a proper human being", to support a family and bring up children. In principle wages are close to the value of labour power; the constant struggle of the workers does not permit capitalists to pay much less than this amount. In South Africa, however, it is declared openly that the value of labour power is not paid by the capitalists. This is a breach of the "normal" economic laws of capitalism itself. It is the result of apartheid.

In spite of all the endeavours and experience of racists, the waves of protest and struggle which have spread round the republic of apartheid in recent years have not passed the gold fields by. The events of September 1973 in Carltonville, at one of the mines of the Anglo-American Corporation, when the police fired on striking miners and killed eleven people caused a great international uproar.

In the last few years some things have begun to change at the gold mines. Even the diehard racists realise that the present system cannot go on forever and that it may lead to an explosion. The flow of fantastically cheap manpower from the neighbouring states has decreased, particularly from Mozambique whose people have cast off the colonial rule. On the other hand, repeated rises in the price of gold and the profits of companies has enabled them to agree to wage rises for workers without great loss. The companies vie with one another to advertise their "achievements" and innovations in the sphere of medical care and training. Blacks are now more often made team-leaders and low-level managers.

But essentially apartheid still remains the basic form of life at the gold mines, and some of its forms are growing harsher instead of easing.

The Destruction of Art

"Numberless goldwares must have been produced in Egypt over a period of 3,000 years, but all that is left are the contents of a dozen or so royal graves, the most important of which all belong to the Middle and the New Kingdoms. Only the innate conservatism of the Egyptian artists and the additional evidence provided by paintings and sculptures make it possible to assess Egyptian goldsmithing on the basis of such slender evidence. And this is a recurring problem. So little is left to us from Ur, so little from Minoan Crete, so little even from Classical and Hellenistic Greece."[27]

This is a quotation from a new book on the history of the use of gold in jewelry-making.

By becoming money and the embodiment of wealth, gold destroyed not only human lives, but also works of art. Ever since ancient times religious and secular ornaments, often of great artistic, historical and cultural interest, have been made of gold (and silver). However, for thousands of years the value of the metal contained in them was more important than their artistic significance. It is fortunate that the tomb of Tutankhamen was not discovered until the 20th century, when finds no longer risk being melted down. But Tutankhamen was an "ordinary" pharaoh, who died when he was still quite young. We can only guess what treasures have perished when the tombs of the great rulers Ramses II or Turthmosis III were robbed. The value of the several kilograms of gold from which the mask of Tutankhamen was made is considerable today as well, but it is nothing compared to the historical and artistic value of this splendid work of Egyptian masters of the 14th century B.C.

The burials of the Egyptian pharaohs and Scythian and Thracian rulers and the tombs of pre-Columbian America have always attracted robbers from ancient times up to the present day. Only in very rare cases has this booty eventually reached museums. More often than not articles of gold and silver were melted down and turned into money. For secular and religious rulers their treasure stores were always reserve funds of precious metals. In cases of extreme and not so extreme need these articles were also melted down into metal.

This phenomenon could be called *gold articide,* by analogy with the word "genocide": in Latin "ars" means "art" and "occidere"—"to kill".

Gold articide is closely linked with conquest and bloodshed, religious fanaticism. Russian chronicles describing the fall of Kiev, Ryazan, Vladimir and other towns in the flames of the Tartar-Mongolian invasion and lamenting the death of thousands of peaceful citizens, do not neglect to mention also that the infidels stole the church plate, seized all the gold they set their eyes on, and tortured their victims to find out where gold had been hidden. This is the main reason why there are so few articles of jewelry by old Russian masters of the pre-Mongol period in our museums.

In 1520 the great German artist Albrecht Dürer saw Aztec treasures from America in the palace of the Spanish vice-regents in Brussels. They included, for example, "a sun of pure gold a whole arm's breadth wide". Dürer writes: "And I have seen nothing in all my days that delighted my heart as much as these things. For I saw among them wonderful, artistic things and I wondered at the subtle ingenuity of people in foreign lands."[28] Another fine master, Benvenuto Cellini, saw these treasures and admired them. These Aztec works of art, including the sun which Dürer described, have disappeared without trace: their owners were not such fine connoisseurs of pagan art and had them melted down. It is only from descriptions as well that we know about the gold interior of the palaces of the Inca capital.

Even from the works of Cellini himself, the famous 16th-century jeweller of the age of the Renaissance, when one would have thought people had learnt to value art, *only one* gold salt-cellar of the French King Francis I has come down to us, and it escaped the common fate by a miracle. This salt-cellar is a complex sculptural composition with the figures of gods and mythological beasts. The material with which the master worked destroyed his creations. In the enlightened 19th century in Pa-

pal Rome works of art were melted down to pay indemnities to Napoleon.

During the period of the gold standard "aesthetic treasure", i.e., articles of gold, was a reserve of money circulation. When economic conditions brought about a reduction in the money supply, part of the gold coins was smelted into more or less crude articles and kept in this form. When conditions changed the gold went back into circulation, for which plate, vases or statuettes were turned into bars and exchanged for coin. This was a spontaneous economic mechanism. But who can guarantee that together with the crude articles, in which the labour of the jeweller was only a small part of the value, some valuable works were not sometimes turned into featureless metal?

Gold articide still happens today. Because of the high cost of the material artistic articles of gold are portable and most suitable for theft, smuggling and concealment. If the thieves, speculators or "collectors" do not melt them down, realising that they are more profitable as works of art than as metal, they often keep them secretly in safes, and at times of great tension bury in the ground.

The theft and smuggling of archaeological treasures is big business in the West today. When archaeology is combined with gold this business becomes even bigger and more profitable.

VI. THE POWER OF GOLD

It is impossible to quote all the statements by writers of all ages and peoples about the power of gold: there is an infinite number of them. The power of gold over people is astounding, it has been cursed and extolled, but no one has remained indifferent to the yellow metal.

The time has come to examine what this power really means, what is its nature and why it has become an element of social consciousness over the centuries. Whence these turbulent emotions? Iron, for example, a far more useful metal, occupies a far more modest place than gold in the people's emotions—if not in their minds.

Gold, Money, Capital

In the pamphlet "The City of the Yellow Devil" by the great Russian writer Maxim Gorky we find the following image: iron as the slave of gold. In this work the writer describes his impressions of New York which he first visited in 1906. Gorky saw this multi-million city as the monstrous product of bourgeois civilisation.

The yellow devil reigns supreme in the terrible, inhuman city, in the capitalist world. People are in its power. Whereas the workers are blind tools, the rich are slaves of the yellow devil. The yellow lights of the advertisements seem to glitter like gold. Towards nightfall "the clang of iron, driven everywhere along the streets by the insatiable goading of Gold, never dies down... The dazzling effulgence of molten Gold pours from the walls of the houses, from shop signs and the windows of restaurants. Insolent, blatant, it triumphs everywhere, making the eyes smart, distorting faces with its cold glitter."

The pursuit of gold, the lust for and power of gold fill people's lives: "It is as if, somewhere in the heart of the city, a huge lump of Gold were spinning at a terrific pace with voluptuous squeals, powdering the streets with the finest particles, which people catch and seek and clutch at eagerly all day long. But evening falls at last, the lump of Gold begins to spin in the opposite direction, raising a cold blazing whirlwind, drawing people into it so that they will give back the gold dust they caught during the day. They always give back more than they got and next morning the lump of Gold has grown larger, it revolves at a swifter pace, and the exultant screech of iron, its slave, the clang of all the forces it has enslaved, sound louder still."[1]

How remote this is from the material for dentistry and electronic contacts! Gold here attains the apocalyptic power of a symbol, a symbol of money or, to be more precise, a symbol of *capital*. For it is capital that enslaves people, that makes people its tools, slaves and servants. The "lump of Gold" growing fatter each day is seen as the symbol of the exploitation of labour by capital, the continuous accumulation of surplus value being converted again into capital.

Some say that New York is the most American city of the USA, others, on the contrary, that it must on no account be considered the face of America. This depends on one's point of view. But Gorky at any rate saw this city as the symbol of American capitalism.

The image of gold was for him a reflection of the real economic fact that gold was the prevalent form of money, a synonym for money. And money has always been the initial and most palpable form of capital. The capitalist must have money, first and foremost, to buy machinery and raw material, to hire workers, and to begin production. By selling the produce he obtains more money than he invested in the business.

Today money and capital are no longer so directly associated with gold, although the essence of capital remains the same and the dimensions of its accumulation have increased many times over. A modern writer would probably find a different image and different words for New York—the city of big capital, backbreaking labour, mass unemployment, racial strife and record crime. But Gorky's Yellow Devil is still understood today just as the writer wished: as a symbol of the inhumanity of capitalism.

For people who see bourgeois civilisation as the pinnacle of creation, the power of gold conjures up quite different associations. For them gold is rather a god, and they sing its praises. In 1943 in Chicago a book came out with a title that speaks for itself: *Gold, the Real Ruler of the World*. The author believed that the abolition of the gold standard in the USA was a temporary setback, and that after the war gold would be restored to its throne. The author's name, Franklin Hobbs, would mean nothing to the reader if the book were not dedicated to Thomas Hobbes, "my illustrious and much maligned ancestor who gave the World The Leviathan".

On the one hand, Mr Hobbs, a descendant of the great 17th-century philosopher, is an obvious crank. The book is furnished with numerous statements and aphorisms set in half-inch high type, such as "Gold is the key to the destiny of nations", "Gold is to commerce what blood is to the body", and "Gold is the cornerstone of the whole business and financial structure". Hobbs takes the Bible quite literally and evidently does not doubt that the world was created a few thousand years ago. "At the very threshold of the World's history, just immediately following the Creation, Gold appeared, and its power began to assert itself." He also writes: "Gold, in the language of the Bible, as interpreted through the ages, meant power; meant, in fact, an instrument of God."

This deification of gold, somewhat strange in the 20th century, is, however, combined with some curious observations and thoughts. Even the writer's misunderstandings are interesting. Let us quote one such passage: "The most maligned, and the most misunderstood and, we had almost said, the most hated inert thing in the world is Gold. It is surely difficult to understand why Writers and Speakers and Politicians and Bankers all get excited when they talk about Gold. They rave and rant and call names, when Gold is mentioned... Just why a mere metal should rouse such a feeling of animosity is hard to fathom."[2]

Gold is not just "a mere metal", of course. Gold is money and capital, it is connected with the property relations between people, with the private interests of individuals and social groups. Here, as Marx put it, the Furies of private interest are involved. It would be naive to expect calmness and impartiality in the discussion of such an object.

Historically gold has been not only the embodiment of capital in general, but the tool and banner of a certain type of capital, a group of the bourgeoisie. At the end of the 19th century the gold question was at the centre of the political life in the USA. The Republican Party, which expressed the interests of industrial and banking capital, favoured the introduction of the full gold standard and the demonetisation of silver. The Democratic Party, which was supported in particular by the farmers, was for silver and bimetallism. The real explanation of this position was that people thought the mass minting of silver on a par with gold would lead to a rise in the level of prices, and this would have benefited the big farmers. The leader of the Democrats, William Jennings Bryan, held his opponents up to shame: "...We will answer their demand for a gold standard by saying to them: You shall not press down upon the brow of labour this crown of thorns, you shall not crucify mankind upon a cross of gold."[3]

These words contain a strong element of demagogy: the true interests of the working people were, of course, connected not with gold or silver, but these interests could not find political expression at that time. At the Presidential elections of 1896 Bryan was defeated. In 1900 a law was passed on the gold standard, and the demonetisation of silver entered its final phase.

It must be said that the association of gold with *big capital*, its representation as the *money of the rich*, have a long history. It is sometimes challenged by silver and copper, and sometimes by paper money, as the "money of the ordinary people". John Law, the famous Scotsman who "invented" paper money, speculated on this at the beginning of the 18th century. He believed that by replacing the gold in circulation with paper money he would ensure the people an abundance of money and, consequently, economic prosperity. This experiment in France ended with the first case of large-scale paper-money inflation in history. Law enriched not the people, but a handful of businessmen, speculators and profiteers.[4]

The idea of the power of gold is by no means the same in different ages. For the Baron in Pushkin's *The Covetous Knight* gold is, so to say, the potential power over people, over their labour that gold can buy. These include the services of such exotic characters as playful nymphs, muses, free genius (that is, poets, musicians and artists) and hired assassins ("bloody evil-doing"). For the feudal lord the power of gold was not yet the power of capital. He found it difficult to part with his gold even to lend it at interest. This is basically the *pre-capitalist* view of the power of gold, it is pure *accumulation* which is similar to present-day primitive hoarding.

In feudal times gold embodied both the wealth and the aspirations of the young bourgeoisie as yet without political rights. In this capacity it is opposed to the land and sword of the feudal lords, that is, landownership and military power. For the bourgeois gold is, first and foremost, money capital, which can and must be put into circulation in order to increase the initial capital. Pushkin's unfinished drama published under the title "Scenes from the Age of Chivalry", contains the following exchange between the monk and alchemist Berthold and the rich bourgeois Martin:

> *Berthold:* I have no need of gold, I seek truth alone.
> *Martin:* The devil take the truth, what I need is gold.[6]

The loan of 150 guldens which he gives Berthold after much wrangling is also an investment of capital for Martin: he believes the alchemist will discover the secret of making gold, then Martin will receive an enormous dividend on his capital.

Capitalism develops and enhances all the aspects of the power of gold. Gold has the ability to purchase any commodity (or service), to settle any debt. Within each country paper-credit money may have the same ability, particularly if it is exchangeable for gold. But on the world market national currencies have an extremely limited circulation, or are not accepted at all. There only gold has the features of a universal means of purchase and payment. With the development of the economy and commerce the range of goods and services that can be paid for in gold is expanding tremendously.

As we know, gold is no longer in circulation and does not perform the functions of money in the domestic economy of capitalist countries. In order to act as a means of purchase and payment, it must be sold for paper-credit money. This makes it akin to other commodities. However, it is hard to find a commodity which has such a guaranteed market for sale and which can be sold as easily as gold. The growing scale of hoarding shows that the role of gold as a universal means of purchase and payment has not been undermined, although it has changed. In present conditions the position of gold in the international monetary system and on the world market has also changed. But today also the gold reserve is an important index of a country's international standing, a measure of its potentialities on the world market.

Under capitalism gold is becoming the absolutely social materialisation of wealth, as it were, the most effective, indisputable and unconditional form of wealth. Any commodity can be regarded as congealed human labour in its abstract form, in a form where the differences between the concrete types of labour disappear, where only its amount is important. But in gold this property of the commodity has reached its limits, because gold presents its glittering essence for the performance of the functions of money.

"Along with the extension of circulation, increases the power of money, that absolutely social form of wealth ever ready for use," says Marx.[6] Wealth in gold and in money was practically the same thing in Marx's time. When Marx talks about the money being "ever ready for use", he is contrasting gold with other commodities which do not possess this property of absolute *liquidity*. In our day it differs in this way not only from other commodities, but also from paper-credit money. The sphere of the latter's operation is limited to the national state, and it has an indestructible tendency towards inflationary depreciation. National currencies that perform the role of international means of payment, the US dollar first and foremost, cannot escape these congenital defects. If we add to this the dollar's "acquired defects", particularly evident in recent years, the advantages of gold as world money become quite clear.

Throughout the last two decades much attention has been paid in the West to the problem of *international liquidity*. This is the problem of the adequateness of reserves of gold and its acceptable substitutes in international circulation to ensure the relatively smooth functioning of world trade and payments. A country's gold and foreign exchange reserves are that part of its national wealth which is in the form of world money, in absolutely liquid form from the point of view of international payments. These reserves may come to a country's aid in the event of a poor harvest or other natural disaster, and help to overcome temporary difficulties in the export sectors of the economy. Many countries consider not without reason that, all other things being equal, dollar reserves are a useful thing, rights to credit from the International Monetary Fund are worth having, but there is nothing more reliable than gold reserves. Its liquidity, its readiness for use is higher than those of any other reserves. Gold remains the pivot, the foundation of every country's international liquidity and retains an important role in the capitalist monetary system as a whole.

In general the role and power of gold should be seen in connection with the economic category and problem of *reserves* and *reserve money capital.* A battle can sometimes be won by a bold blow, but a war, as we know, is won by those who have greater reserves and whose reserves are the most mobile and effective. This is also true of the activity of the capitalist firm and the economic strategy of the state. In order to ensure the successful turnover of all capital, some part of it must be kept in reserve.

A country's gold stock is a national reserve money capital. Of course, it would be wrong to think that the power of a state is determined by the size of its gold stock. The foundation of this power is its economy, production, skilled manpower, and scientific and technological potenial. Natural resources and accumulated reserves of the main types of raw material and fuel may be incomparably more important than piles of gold bars in bank vaults. But history knows of cases when the working of a country's whole vast economic machine has depended on the existence and use of these bars. This role of gold is still not exhausted even today. The instability of the economy and finances of the capitalist countries may increase it.

A Fetish

A fetish is an object said to be endowed with supernatural properties, an object of blind worship. All that we know about gold confirms that it is a fetish. But why and how did it become one?

René Sédillot explains the matter as follows. People's worship of gold is akin to their worship of the forces of nature. It is in no way connected with its social functions. "Ever since man got to know gold, he has been fascinated by it. . . We must conclude that its beauty is captivating, because everywhere from the very beginning man has been captivated by it. The *alpha* and *omega* of the history of gold lies in this statement: gold is a magic metal... Where geologists would like to see a mineral, and economists a commodity, the sociologist and the historian must see a faith."[7]

He then presents extensive material on the role which gold has played in various religions. We find that in some way or other it has been deified by almost all religions, except the Chinese. Incidentally, this too is connected with a superstition: the ancient Chinese associated gold with evil underground spirits.

Does this historical excursion explain the reason for the deification of gold, however? To my mind it does not. The author himself has fallen victim to a kind of fetishism. The "magic" properties of gold are declared to be inherent in it and therefore inexplicable.

There may be a grain of reason in the comparison of gold with the natural forces deified in primitive religions. Its association with the sun is very ancient. We find it in the Egyptian religion and it is retained in alchemy right up to recent times. But this alone can hardly explain the fetishisation of gold. For the gold fetish reached its height not in primitive societies, but at a fairly advanced stage of development, when purely religious, to say nothing of totemic ideas, no longer played such a large role in social consciousness. In the period of the conquest of America it transpired that gold was a fetish not among the naive Indians, but among the enlightened Christian Europeans.

The fetishisation of gold is primarily *social* in nature and it is determined above all by the social, not the natural, functions of gold. While not denying the probable connection of the religious worship of gold with the worship of natural forces, we must view this worship mainly as the reflection in the religious consciousness of its special *social* status.

There can hardly be any doubt that the great role played by gold in the Old Testament is merely, so to say, a reverse projection into mythology, into legendary and semi-legendary history —the projection of the real commodity-money relations of the age when it was compiled, that is, approximately at the end of the 2nd millennium and the beginning of the first millennium B.C. and during the following centuries. At this time gold undoubtedly played an important social and economic role in Palestine.

Franklyn Hobbs, mentioned above, has calculated that gold is mentioned 415 times in the Old Testament. The very first reference throws light on the "magic" of gold. In Chapter II of the Book of Genesis there is a description of the Garden of Eden in which God has placed man just created by him. Adam has already been created, but the creation of Eve has yet to come. Here it is written: "And a river went out of Eden to water the garden; and from thence it was parted, and became four heads. The name of the first is Pishon: that is it which compasseth the whole land of Havilah, where there is gold; and the gold of that land is good."[8] What real geographical facts are referred to in this passage has not yet been fully established. Some specialists believe that the land of Havilah is the central part of the Arabian Peninsula in what is today Saudi Arabia, where gold was actually mined in ancient times.[9] True, there is no suitable river there, but can one demand geographical accuracy from the author of the Bible? And in any case this is of no importance to us. What is important is that gold appears here only as a mineral, an important one, of course, not from the point of view of the naked Adam, but from that of the author who was already well aware of its value.

About the wealth of Abraham it is written: "And Abram was very rich in cattle, in silver, and in gold."[10] There is no trace of magic here, but there is most interesting evidence of the role of precious metals as a special form of wealth in ancient Hebrew society. Although gold and silver are not yet money here in the full sense, they are already very close to it.

We shall not tire the reader by quoting the remaining 413 passages, but shall merely state that the Bible records not the divine origin of gold, but its social status.

In the ornament of temples of many religions, Christianity included, gold plays an important part. According to Sédillot this is because it is itself an element of cult in each of these religions. But another line of reasoning is more realistic: the lustrous reputation of gold arises outside religion, in socio-economic reality, "on the market". Gold is valuable not because it is sanctified by religion, but, on the contrary, religion adopts it because it is valuable.

Thus, the nature of the gold fetish is social. Therefore gold fetishism does not decrease with the growth of civilisation and education, but increases, assuming the most developed, refined forms. The material provided by nature acquires "super-natural" properties—it becomes money. Money is the product of a social system in which people relate to one another not as individuals, but as producers and owners of certain commodities.

This type of relation between people is strengthened with the development of capitalism. The production of goods for oneself, for one's own consumption, becomes a rarity. Everything is produced for exchange only, for the market, for money. Man's dependence on other people has always been an essential feature of human society. But this dependence takes on a market, monetary form only at a certain stage in the development of society. At the early stages of capitalism a person's existence depends primarily on whether he can sell the commodity he is producing. Later on the main thing for an ever increasing number of people is the sale of their labour power, their ability to work. The outside world, the world of other people, confronts man in the form of a featureless mass of commodities and is increasingly concentrated in the form of money, gold, for which these commodities are sold.

In human consciousness gold becomes the abstraction of all commodities, their plenipotential representative, their ruler in a certain sense. It is omnipotent because it can buy everything. People think that this has always been so and must be so, that this state of affairs is not the result of a certain level and form of the development of society, but a law of nature. Hence it follows that nature itself creates gold as money, that gold is predestined to rule human fates. From here it is but a step to deification. In this connection Marx says: "These objects, gold and silver, just as they come out of the bowels of the earth, are forthwith the direct incarnation of all human labour. Hence the magic of money."[11]

An example of this fetishism is to be found in the above-mentioned Hobbs: "Gold always has been money and always will be money. This resulted from natural law and will not be changed by man-made law."[12] The advantage of this statement is its simplicity. It compares favourably in this with the usually rather flowery and imprecise manner of expression of most economists. In another passage Hobbs quotes the aphorism that gold is "the Prince of metals and the King of commodities".[13]

Under capitalism money traverses the path from gold coin to entries in bank accounts, and now often takes the physical form of signals in the memory of computers. It looks as if payments will be made increasingly by "pressing a button" or, rather, by selecting different codes understood by computers. But this cannot change the socio-economic nature of money under capitalism. It expresses basically the same social relations, first and foremost, the relations between capital and wage labour. Money not only remains the pillar of capitalist civilisation, but its power and influence on people's lives are inevitably growing stronger. A punched card or magnetic tape may be a fetish no less than a gold coin, if they embody money and its movement.

But, as we have seen, modern capitalism cannot get rid of the gold fetish in the most literal sense of the word. Even when gold ceases to function as money and, it would seem, returns to the world of ordinary commodities, it retains such specific social properties that it remains an object of greedy adoration and worship.

Hymns and Curses

From time immemorial gold has been associated in human consciousness with god and the devil, greatness and baseness, perfection and monstrosity, nobility and crime. This dual moral evaluation has come down to us in the form of arguments between economists in the West on whether gold is good as money and whether it should be money or not.

Sédillot, in whose book sober sociological analysis alternates with involuntary admiration of this enigmatic, inexhaustible metal, informs the reader: "I do not carry gold, in my heart, on my finger, or in my purse. It is materially absent from my life..."[14] Paul Einzig, however, writes: "I myself have known several very vocal opponents of the monetary role of gold who possessed gold watches, cigarette cases or other objects, even though stainless steel or other metals would have served their practical purposes perfectly well."[15]

These paradoxical statements simply show that human psychology is not all that simple. But do we expect the reader to assume that a person writing a book about gold is bound to worship it in his everyday life? Or that a person who opposes the gold standard should despise this inanimate object?

People have always admired the beauty of gold and always will. But there is much beauty in nature, and the merits of gold do not outweigh everything else that nature has created. It was, of course, not the natural properties of gold that Columbus had in mind in the above-quoted letter when he called gold *perfection*. And it is not for its "glitter" that Balzac's Gobseck sings its praises. The metal's natural lustre increases ten-fold when people look on it as real or potential money. With palpable force gold embodies all the aspects of money, first and foremost, money as the "larva" of capital—value capable of producing new value.

"Modern society," Marx says, "which, soon after its birth, pulled Plutus by the hair of his head from the bowels of the earth, greets gold as its Holy Grail, as the glittering incarnation of the very principle of its own life."[16] As we can see from Marx's remarks, in his reference to Plutus (the god of the underworld and owner of its wealth) he had in mind the following phrase by the Greek writer Athenaeus: "Greed hoped to pull Plutus himself by the hair of his head from the bowels of the earth."

The worship of gold which we find in the customs and literature of many peoples may have various shades. The gold fetish may present different aspects.

There is the motif of the *omnipotence* of gold, its universal capacity to buy everything—commodities, services, even people, their talent, labour and conscience. There is the motif of the *enigmatic*, mysterious nature of gold, amazement that a thing possesses so much power. Gold *levels* people of all estates and status, dividing them according to one principle only: who has how much of it. Gold *stimulates* human labour, energy and enterprise. It vitalises, accelerates and drives ahead. Finally, there is the motif of admiration at the *organisation* of an economic machine in which gold spontaneously, and partly on agreement between people, occupies pride of place, is the main form of money. This form of the gold fetish is, perhaps, the most deep-rooted and widespread among economists.

During the period of the gold standard, up to World War I,

the merits of gold as the basis of the monetary system were usually taken for granted, as something obvious and not requiring substantiation. It was, perhaps, the collapse of the gold standard that made economists and sociologists take a closer look at their old friend and try to comprehend its role in capitalist economy and bourgeois civilisation.

Enough about the primitive, naive forms of gold worship. Let us now examine the themes developed by the academic "gold troubadours" who claim to analyse the problem scientifically.

One of the favourite themes is gold and "free enterprise". Historically the predominance of gold as the basis of the monetary system is associated with the age of capitalist free competition, and we frequently find nostalgia for this world "of order and harmony" in the West. This is what Einzig has to say: "It was substantially correct to say that the period of the pre-1914 gold standard had been (in more than one sense) the golden age of mankind."[17] In this connection it might be apt to quote another of Felix Krivin's paradoxical remarks: "*The golden age.* The well-known iron age, which could have become the age of gold, if not for its excessive passion for gold."[18]

Today it is the supporters of private enterprise and the critics of "excessive" intervention by the state in economic life who urge a return to the gold standard. This breed of individual is very hardy and is found in many spheres, including university departments.

Closely connected with this is another theme—the stability of gold and gold money, their anti-inflationary nature. Capital investment in gold is an "island of stability" in the stormy sea of inflation. But another point is even more important: if gold is money, even to some degree, the economy and the state itself are forced to submit to "gold discipline", and this is a certain guarantee against inflation. The purchasing power of gold forms spontaneously and tends to be stable for a long time. The purchasing power of paper money depends on how much of it is issued by the state. This framing of the question has inclined many, of course, to favour gold as money. No lesser a person than Bernard Shaw wrote with a slight touch of his inimitable irony: "You have to choose (as a voter) between trusting to the natural stability of gold and the natural stability of the honesty and intelligence of the members of the Government. And, with due respect for these gentlemen, I advise you, as long as the Capitalist system lasts, to vote for gold."[19]

Ever since ancient times there has been a kind of division of labour between silver and gold. Silver was primarily internal money and rarely ventured beyond the bounds of the state. Gold has for centuries been world money used in international settlements. The "gold-troubadours" sing the praises of its cosmopolitan nature, its connecting and uniting role on the world market. Incidentally, this does not prevent some of them from extolling the accumulation of large state gold reserves, which always serve as an instrument of nationalism. With naive simplicity Hobbs says of the US gold reserve during World War II: "The great stock of Monetary Gold of which this Country is presently possessed holds within its yellow mass the destiny of nations and of all the peoples of the world."[20] It goes without saying that this belief in the omnipotence of gold turned out to be an empty, albeit harmful, illusion.

The tradition of condemning gold, hurling abuse and curses at this inanimate object, is just as old as that of singing its praises. In the antique world gold (not gold as such, in fact, but the development of commodity-money relations) caused the breakdown of the patriarchal way of life, undermined the old moral values and brought about a "decline in morals". As Plutarch tells us, the semi-legendary law-giver in Ancient Sparta, Lycurgus, tried to combat this evil by forbidding money made of precious metals. Naturally nothing came of this. Sophocles says in *Antigone* that money is the curse of mankind, that it corrupts men's hearts and teaches them to perform dishonest acts.

Various historical sources tell us about the "fear of gold" in Ancient China and among the Germanic tribes in Roman times. This fear is akin to the fears expressed by Lycurgus. We find similar things not only in remote antiquity. When in the 1850s information was received about the gold deposits in southern Africa, the Parliament of the Transvaal Boers—a small people with patriarchal customs—issued special laws prohibiting mineral prospecting and the disclosure of information about the country's mineral wealth.[21]

Ever since Plato the abolition of money has been one of the favourite ideas of utopians who have presented plans for the reorganisation of society. They wanted to destroy capital and the exploitation of man by man indissolubly linked with it. But the utopian critics of bourgeois society saw capital in its glittering golden dress just as those who sang its praises did.

According to some scholars,[22] the following lines by Shakespeare in the tragedy *Timon of Athens* were written under the influence of Thomas More, the author of *Utopia*, the work which provided the name for the whole school of thought.

> *This yellow slave*
> *Will knit and break religions, bless th' accurs'd,*
> *Make the hoar leprosy ador'd, place thieves,*
> *And give them title, knee and approbation*
> *With senators on the bench.*[23]

Shakespeare calls gold the common whore of mankind. Pierre Boisguillebert (1646-1714), the eminent French thinker and one of the founders of the science of political economy, wrote that precious metals as money had acquired a tyrannical power over people and become the main cause of economic disasters. Since money is not in itself an important object of consumption, he regards it as something artificial and unnecessary in rational human society. He criticises money in strong terms: "Corrupt hearts have erected . . . gold and silver . . . into idols. They have been turned into gods to whom more possessions, valuables and even people have been and still are being sacrificed than blind antiquity ever sacrificed to these deities, which have long formed the only cult and only religion of most peoples."[24]

The Russian democrat and humanist A. N. Radishchev, describing the life of Lomonosov, an outstanding Russian scientist, and his interest in minerals, makes the following remark about precious metals as money: "Gold and silver, which are most precious metals in their perfection and which did formerly serve as ornament, have now been transformed into tokens that represent all manner of money-grabbing. And only then in truth, then was the human heart seized with an insatiable and base passion for riches, which, like an all-consuming flame, doth grow stronger as it receives sustenance. Then, leaving his original simplicity and his natural occupation of tilling the soil, man did commit his life to the stormy waves, or, indifferent to the hunger and heat of deserts, did traverse them into unknown lands to win riches and treasure. Then, spurning the sun's light, he did descend into the grave and, tearing asunder the bowels of the earth, did burrow himself a hole, like a reptile seeking food in the night."[25]

In this archaic, but powerful language Radishchev was expressing an interesting idea. He had noticed the dual role of precious metals in the development of mankind: on the one hand, they stimulate people to make long journeys and study the earth's riches, thereby promoting progress, and, on the other, the vileness of a society in which everything is subordinated to the pursuit of money derives from them.

The illusion that all the evil, all the vices of capitalism come from money and, more particularly, from the predominant role of precious metals, is extremely persistent. There are countless projects to free capitalism from its vices by abolishing money. This was the cherished idea of Pierre Joseph Proudhon, one of the forerunners of scientific socialism. Proudhon, who hated and bitterly denounced capitalism, assumed that one need only introduce direct non-monetary exchange of commodities according to their labour value (that is, to the expenditure of labour on their production) to be established and verified in a certain way, and exploitation, injustice and poverty would disappear as a matter of course. Gold and silver would return to their "natural" status of ordinary commodities. The socialism of Proudhon and his followers was petty-bourgeois: they wanted to improve capitalism, not to abolish it, and they did not set the task of abolishing private ownership of the means of production.

In the 20th century, particularly its latter half, the direct identification of gold and capital, the direct association of gold with the vices of capitalism, are gradually losing their meaning. This is due to the abolition of the gold standard and the partial removal of gold from the monetary sphere. The petty-bourgeois reformer's hatred is now more often directed against the gigantic monopoly bank than against gold. Whereas in the 19th century criticism of gold was also criticism of capitalism, today it is quite impossible to imagine a socialist of any kind for whom the fight against the omnipotence of gold would be an important part of his anti-capitalist programme.

Criticism of gold is now perfectly respectable and is carried on from the positions of state-monopoly capitalism. Impassioned cursing of gold is now giving way to sober arguments that it should "know its place", that it must not be allowed to restrict state regulation of the economy.

The opponents and critics of gold call the gold standard and the monetary functions of gold in general a "barbarous relic". This term, coined by John Maynard Keynes in his book *A Tract on Monetary Reform* (1924), has become extremely well known and turned into a kind of slogan or curse.[26] Gold is now blamed primarily for the inflexibility of the monetary system based on it and the difficulty of conscious regulation of this system by the state.

The supporters of gold maintain that only gold is capable of saving capitalism from inflation, monetary crises and other upheavals. Naturally, both sides wish capitalism well, but they are searching for different ways of achieving this.

The debate sometimes becomes extremely heated reminding us again and again of the fact that no one can remain indifferent to gold. Both sides shower barbs of sarcasm on their opponents. The critics of gold represent its supporters as hopeless retrogrades, upholders of obsolete dogmas. Determined not to be outdone the latter maintain that the opponents of gold are egg-heads, academics cut off from the real world, ready to take up any fashionable theory, and their recent students, now ambitious, extravagant officials in the higher echelons of the state. One author says that the dividing line is age: the old men are in favour of gold and the young against it. There may be some truth in this argument, but age is not a really important thing: the opponents of gold are usually supporters of active state regulation of the economy, followers of Keynes; those in favour of it put their faith in the market mechanism. Both positions also have many shades. Some are for total and immediate demonetisation, others for gradual and cautious removal of gold. Some are for restoring the gold standard in its full, original form (they are few in number), others for partial and limited restoration, usually only in the international sphere. After all a specialist's standing and authority demand that he adopt a position somewhat different from that of others who share his views.

The Magnates

For centuries people have associated a rich man with a bag of gold. However, in fact only the collector of treasure, a figure more typical of the Middle Ages than of "enlightened" capitalism, keeps most of his wealth in gold. With the development of this system the desire to keep as little of one's capital as possible in gold, and money in general (including bank accounts), grows stronger. For gold as capital is dead. It not only does not yield profit, but also requires expenditure on storage, insurance, etc. Ready cash also yields nothing. A bank account at best yields interest, but at a lower rate than if capital is invested in an enterprise. Capital by its very nature cannot lie idle. If today, as we have seen, hoarding and capital investment in gold are in vogue, this is explained not by the natural preciousness of gold, but by upheavals in the economy and the socio-political system of capitalism—inflation, instability and crises.

But at the same time the money form of capital is of great economic importance. In the course of the circuit of capital a certain part of it must always be in the form of money for effecting the payments due, and also as a reserve in case of unforeseen circumstances. In a period of economic difficulties an enterprise with ready, mobilised money capital has a better chance of surviving. With the development of capitalism there is a rise in the importance of banks which hold practically all their capital in money form. The following point is also important. Wealth is usually perceived, first and foremost, as its tangible and regular result—income, and income as a rule is obtained in the form of a certain sum of money. It is wrong to think that wealth and capital are invariably money, but money is a means of measuring wealth, an essential and vitally important part of it, its visual expression. Therefore, when the image of a money-bag is applied to a rich man, it is not only a figurative turn of phrase, but also an economic reality.

Naturally, the wealth of the rich takes the concrete money form which is predominant in the society in question. When cattle were money, the owner of large herds was considered a rich man. Today a rich man is, first and foremost, the owner of large blocks of shares, other securities and bank accounts. But gold, as we have seen, was "already almost money" in the time of Abraham, was an important and major form of money for many centuries, and retains features of money today. This is why the association of wealth with gold is still alive in people's minds and language today.

The connection between wealth and gold has another specific meaning. The period of the gold standard (late 19th-early 20th centuries) was the classic age of the rich, and not simply of the rich, but of the magnates of capital, the creators of financial and industrial empires and vast family fortunes. Nothing prevented them from getting rich and leaving their wealth to their heirs. The state had no need to intervene, and hardly ever did, in matters of private capital. Income taxes and death duties scarcely existed. Money was stable, and there seemed nothing more secure in the world than British Consoles (state bonds that yielded a steady income). The world was divided up between the imperialist powers, and national liberation movements did not yet pose a serious threat to foreign companies. The frontiers between states did not hinder them, and for the first time in the history of capitalism international monopolies were formed. This was a world in which inequality between people, classes and nations was considered natural among the bourgeoisie and bourgeois intellectuals. Wealth enjoyed respect and aroused envy.

True, a strong working-class movement was already in existence, and there were influential Social-Democratic parties in many European countries. The revolution of 1905-1907 in Russia was a stern warning not only to autocracy but to world imperialism as well. There was unrest in China, the Balkans, and other parts of the world. But this did not yet seriously frighten the bourgeoisie of the main capitalist countries.

The symbol of this period, this world of the owners and magnates of capital, was gold. For them it was indeed a golden age. The metal was not the object of panicky speculations and fearful hoarding that it is today. It was the official basis of monetary systems and circulated in the form of gold coin.

Gold money has always been associated in the West and still is with financial and political conservatism, with the struggle against reform. Although this is by no means a general rule, the people who are "for gold" are often those who oppose trade unions, taxation of high incomes, universal suffrage and so on.

In Arthur Hailey's novel *The Moneychangers* mentioned above Margot Bracken, a woman lawyer with radical views, says in an argument with the financial conservative D'Orsay who longs for a gold currency: "On a gold standard, even fewer people than now would own most of the world's wealth, with the rest of mankind left bare-assed."[27]

Incidentally, D'Orsay's views and language were borrowed by the writer, who, as we know, is distinguished for his scrupulous study of the background for his novels from real sources. The prototype could, for example, be Franz Pick, the publisher of financial reviews and reference books. He furiously attacks professors who preach the demonetisation of gold, the "absolutely immature bank presidents", the "amateurs in the American Treasury", and the "even less intelligent officials of the Cabinet", all those who indulge in malicious inventions about the obsolete nature of the gold standard and pursue an "anti-gold" policy.[28] Both the imaginary D'Orsay and the very real Franz Pick write and publish their bulletins for very wealthy people and play on the understandable nostalgia of the rich for the good old days and customs of the gold standard.

Thus, rich men and gold, magnates and gold, these concepts are very close to each other in legend and real life. But this closeness becomes particularly concrete when we are talking of gold magnates, the leaders of big capital, in whose hands the mining, marketing and processing of gold lie.

Thanks to the efforts of hired hacks and conscientious historiographers an extensive literature has emerged on the founders of the South African gold industry. These times are portrayed as an heroic and romantic age, and the founders themselves as the embodiment of energy, enterprise and business sense.

Cecil Rhodes (1853-1902) is usually regarded as the first of the "gold heroes". The name of this English coloniser was given to the country of Rhodesia, but has now disappeared from the political map of the world, being replaced by the old African name of Zimbabwe. Rhodes' wealth was founded on diamond mines discovered ten years before the Witwatersrand gold belt. In 1887 Rhodes and his companions set up the Gold Fields of South Africa Ltd., which still mines about 20 per cent of all South African gold. Rhodes held the promoter shares and was the main administrative official in the company, which gave him a total annual income of £300,000 to £400,000. He later gave up his special promoter shares, receiving in return ordinary shares to the value of £ 1,300,000.[29] Rhodes was the Prime Minister of the British Cape colony and did a great deal to provoke the Boer War which devastated southern Africa.

Even more dramatic was the career of another gold king, Barney Isaacs, who changed his surname to Barnato (1853-1897). The son of a London East End shopkeeper, this buffoon and poor adventure-seeker made his fortune on diamonds, like Rhodes. When at the height of the gold rush in 1888 he arrived in Johannesburg and was asked what he was going to buy, he replied, "Everything".[30] Six months later Barnato was already a large owner of gold-bearing land. In partnership with his brother and nephew he set up the Johannesburg Consolidated Investment Company, which is still flourishing today. The London Stock Exchange avidly bought up the Barnato company shares. The little buffoon from Whitechapel returned to London as a financial magnate, and the Lord Mayor held a banquet for him. Barnato's eventful life ended in madness and suicide.

The name of Ernest Oppenheimer (1880-1957) has already been mentioned. He pioneered the drawing of American capital into the gold-mining industry of South Africa. This is reflected even in the name of the concern founded by him in 1917, the Anglo-American Corporation. A South African historian of Witwatersrand says of him: "Certainly luck was with him. But there were also courage and imagination, without which he could not have won. Indeed imagination, which the bankers call 'foresight', was his supreme quality. He had the gift of being able to take the bare details of a plan and build in his mind's eye the complete structure that would be there five or ten years later."[31]

The son of Sir Ernest, Harry Oppenheimer, is still managing the family empire today. But times have changed. It is many years since the gold-mining industry had sensational discoveries and feverishly set up new concerns. In the management of the company the role of specialists—administrators, engineers and financial experts—is now more evident. The mergers and swallowing up of companies and the appearance of new firms and persons are more likely in the oil or engineering industries.

Nevertheless gold remains a special commodity. It cannot be

said that it is more important than oil: it is just economically different from oil or any other type of mineral. To quote Eric Chanel: "Gold remains a superior form of money, not subject, it would seem at least, to the caprices of the day and people. Everyone therefore turns to it in confusion when the national or international economic and political situation deteriorates. This is the reason why it plays the role of a barometer or indicator of tension in economic systems, particularly in the Western capitalist system, of which it is one of the symbols."[32]

Thus, we have returned to the thesis that gold is a symbol of capitalism. It is not only a symbol, however, but also an important economic category. The following chapters deal with the economic role of gold in modern capitalism and the evolution of that role.

VII. THE GOLD STANDARD AND ITS AGONY

The classical political economist Adam Smith, in beginning to outline the problem of value in his famous book *The Wealth of Nations* (1776), apologises to the reader for the difficulty of the material. He says he will endeavour to explain value "as fully and distinctly as I can . . . for which I must very earnestly entreat both the patience and attention of the reader: his patience, in order to examine a detail which may, perhaps, in some places, appear unnecessarily tedious; and his attention, in order to understand what may perhaps, after the fullest explication which I am capable of giving, appear still in some degree obscure."[1]

Turning to the question of the role of gold in the capitalist monetary system, we should like to say something similar, particularly because this subject is close to Smith's: for gold as money is first and foremost a measure of value.

One would expect the gold standard to be simpler than the atom. But here we come up against the difficulties which distinguish the social sciences in general from the natural sciences. In this sphere experiment and the experimental confirmation or refutation of theories are impossible. Social phenomena change with time, and what was true a few decades ago may prove untrue today. The revealing of causal relations is greatly complicated by the incredible variety of operating factors, tendencies and counter-tendencies. Factual and statistical data are not sufficiently reliable, and often do not exist at all.

Therefore, in spite of the vastness of the specialist literature on gold as money and on the gold standard, much in this sphere remains disputable. There is no agreement even on the essence of the gold standard, on the time limits of its operation and the causes of its collapse, to say nothing of the present and future status of gold. Emotions here often take the place of facts, and mutual accusations that of arguments.

What is set out on the following pages is merely a highly popular and simplified introduction to the problems of the capitalist money and of the international monetary system.

What Was It Like

The period of the gold standard is rather vague at both the beginning and the end. In the broad interpretation the beginning of this period is dated back to the 18th century or at least to 1816-1821, when the exchangeability of banknotes for gold was restored in Great Britain after the Napoleonic wars. But in the other large countries, alongside gold, silver still remained the basis of the monetary system or even operated alone (in Russia, for example). Gold was not yet the only form of world money. This state of affairs did not develop until the 1860s. As Galbraith writes, juridically gold was acknowledged as the only form of world money at a "relatively uncelebrated conference" in Paris in 1867.[2] Many authors date the beginning of the complete ("classic") international gold standard to somewhere around this date, although silver fought rear-guard battles until the end of the century, even later, and in the internal circulation of many countries (particularly in Asia) has not been ousted by gold even now.

Nor are things so simple with the end of the period of the gold standard either. As a complete and developed system it ceased to exist as early as 1914, but the USA kept the gold standard, and Britain, France and other countries restored it partially after World War I. It suffered its second "collapse" in the 1930s, in the period of the deepest world economic crisis when the exchangeability of banknotes for gold was abolished everywhere. But the international monetary system (the so-called Bretton Woods system) that developed after World War II retained a very important role for gold. The "final end" of the gold standard is now usually regarded as 1971, when the international convertibility of the dollar for gold was abolished, thereby breaking the last thread that officially connected the world of paper-credit money with gold.

Thus, the period of the gold standard covers about two centuries at most and half a century at least. But the whole point is that it was the monetary system that corresponded to a certain stage in the development of capitalism, the stage of *industrial (pre-monopoly) capitalism.* We sometimes call this stage "free competition capitalism". In the late 19th and early 20th century it was replaced by the stage of monopoly capitalism. The gold standard continued by inertia, as it were, and even spread in the first decade of the 20th century. The whole subsequent period, from World War I and the socialist revolution in Russia, the age of the general crisis of capitalism is marked by the gradual decline of the gold standard. It does not correspond to the conditions and needs of modern capitalist economy.

At the beginning of the present century many observers thought that an ideal monetary system had at last been set up, both inside the main countries and internationally. Its simplicity, smooth functioning, automatism and independence of the arbitrary actions of officials aroused the admiration not only of the paid hirelings of capital. It seemed beautiful, like a mathematical construction or a well-running machine. This beauty was not without flaws, but more about that later. As usually happens in life, it was short-lived. The gold standard was merely an episode in the history of capitalism, although a highly notable episode. Before proceeding further, let us note three facts which are of special importance for a correct understanding of this episode.

First, the short duration of the gold standard does not exclude the fact that gold was money before its introduction and preserved the functions of money in changed form after its abolition.

Second, during this period gold was by no means the only or even quantitatively the predominant form of money; under its aegis there was a rapid development in the different types of paper-credit money.

Third, the gold standard, even at its height, existed in its complete form only in a group of industrially developed countries; the rest had to adapt as best they could to the system established by the powers that be.

The system of the gold standard included some basic elements. They were not all present in all countries at all times, but there was an undoubted tendency towards their dissemination.

Gold coin was in circulation. It was the main form of money, in relation to which paper-credit money and money made of other metals played a functionally subordinate role. According to statistics, in 1913 the three main countries (the USA, Britain and France) accounted for 62 per cent of the world gold circulation and this proportion was growing. The quantity of gold coin in Germany was also significant. There was relatively little gold coin in circulation in Russia—two or three times less in value than in each of the countries mentioned.

All other types of money were exchangeable for gold at face value. The first line of this exchangeability was created by the private commercial banks which always held coin reserves. In the final analysis exchangeability was guaranteed by the central banks which held the national reserves of gold. The three above-mentioned countries accounted for 51 per cent of world central reserves in 1913. As we can see, this percentage is lower than their percentage in coin circulation. The fact is that Britain, which was the centre of the international monetary system, traditionally held a very modest gold reserve. The USA already possessed the largest gold reserve (1,900 tons), with Russia in second place (1,200 tons) and France in third (1,000 tons).

The central gold reserves were the *main reserve of world money, of international means of payment.* In 1913 the gold reserves of all the countries in the world were valued at 6.8 billion dollars, and the foreign exchange reserves at only 800 million dollars. But here the dollar is used only as a unit of account: the central banks had hardly any real reserves in dollars (that is, foreign deposits in US banks and American government securities held abroad), and almost all reserves were in pounds sterling and kept in British banks.

The gold content of each monetary unit was determined by law and maintained unchanged, as a rule, for decades and even centuries. Because of this the relationships between monetary units, exchange parities, were also stable. The gold content of the pound was 7.322 grammes of pure gold, and of the dollar 1.505 grammes, the parity of the pound in dollars, as the quotient resulting from the division of the former by the latter, was

4.87. Equally stable relationships linked the pound and the dollar with the French franc, the German mark and other major currencies.

Free *mutual currency convertibility* (the free exchange of one currency for another) existed. The exchange took place at market rates, which deviated in very narrow bounds from parities, usually not more than one per cent either way. The rates could not vary any more, because a debtor who was obliged to settle a debt in foreign currency always had the choice of buying that currency with his own national currency or buying and transferring gold. For example, over the period 1889-1908 the rate of the pound sterling in Paris (also called the "rate on London") differed not more than 0.67 per cent from parity. The dollar rate in Paris (the "rate on New York") deviated slightly more from parity—up to 1.18 per cent. This difference was explained by the fact that the transporting of gold between Paris and New York was more expensive than between Paris and London.

There was free exporting and importing of gold, and gold was constantly being transferred between countries. This does not mean, of course, that all foreign trading and other international transactions were paid for in gold. The vast majority of payments were made by a currency transfer between banks. Gold came into play only in extreme cases, as the last means of settling the balance of payments deficit, that is, the remainder of all the mutual claims and obligations between countries. For example, the international payment turnover for 1894 is estimated at 20 billion dollars, whereas the flow of gold between countries was about 700 million dollars. Each dollar in gold "turned over" 30 dollars worth of commodities and services.[3]

The normal operation of this system depended on the gold reserves of the central banks. In order to ensure the smooth exchangeability of banknotes laws were passed which obliged the central banks always to have an amount of gold representing not less than a certain proportion of the sum of banknotes issued by them. In order to keep gold and not to refuse to exchange banknotes when there was a great demand for the metal, banks had two main devices. First, it was possible to obtain credit abroad in foreign currency and, if necessary, exchange it for gold. Second, the banks pursued a restrictive credit policy, primarily by raising the rate of interest on their loans to private commercial banks in order to reduce the amount of credit and the circulation of banknotes. With a high rate of interest commercial banks take fewer loans from the central bank and themselves grant fewer loans to industrialists, tradesmen and the public. Money becomes "dearer" and there is less of it. As a result the demand for gold from the central bank may also drop. Moreover, high rates of interest attract foreign capital into the country, and this improves the balance of payments and reduces the need for gold for settling the trade deficit, for payments abroad.

This, put simply, is the gold standard. Here, however, we have been dealing only with its main features, external manifestations, its technical aspect, so to say. More important is the question of the economic mechanism of the gold standard.

The Mechanism

In theory this is similar to the system of communicating vessels. Like the liquid in these vessels, the gold in a state of equilibrium is distributed between countries roughly in accordance with the needs of the economy of each of them. If the equilibrium is disturbed, either as a result of an increase in the output of gold in one of them, or of sharp changes in the balance of payments, forces arise spontaneously that after a certain time restore the normal position again.

This normal position, or state of equilibrium, means that the balances of payments are satisfactorily balanced and, also, that commodity price levels in the various countries are more or less the same. The latter condition should be understood as follows: one pound sterling can, for example, purchase about as many commodities as 5 dollars in the USA (let us recall that this was roughly the parity between them) and a corresponding number of marks, francs, etc.

What kind of mechanism can restore and support this equilibrium? Let us illustrate it by an example.

Let us assume that for some reason Britain's balance of payments with the rest of the world (we shall regard it conventionally as a single entity) had deteriorated. The exchange rate of the pound sterling has dropped and it has become more profitable to exchange pounds for gold and transfer it abroad. There is an outflow of gold and the gold reserve of the Bank of England is dropping. In the rest of the world, on the contrary, there is more gold.

This produces the following symmetrical processes. Gold is the basis of bank credit in the country. When it leaves Britain the reserves of the banks shrink and they cut down their loans to industry and commerce. As a result of this, all other things being equal, the demand for commodities falls and there is a drop in prices. In the rest of the world, conversely, the gold inflow becomes the basis for extended bank credit, a rise in demand and an increase in prices.

More debatable is the presence of another hypothetical mechanism of the symmetrical price changes, which is connected with the *quantity theory of money*. According to this view which goes back to David Ricardo, the outflow of gold brings about directly and immediately a drop in prices, because there is less money in the country. A reduced amount of money is confronted by the former mass of commodities, and the price of the unit of each commodity is bound to fall. In the countries to which the gold is flowing the opposite takes place. Let us remember that here we are discussing gold money or money exchangeable for gold. Marxist economic theory does not accept this explanation and does not agree with the quantity theory of money. Why? Because it is not true that the value of gold is fixed in circulation independently of the conditions of *production* of commodities and gold. This theory gives an extremely one-sided interpretation of why prices rise and fall, connecting both with the amount of money alone.

Be that as it may, in the country from which there is an outflow of gold, prices tend to drop, and in the country to which the gold flows, they tend to rise. Common sense tells us that commodities must tend to flow from a cheap market to an expensive one. Therefore Britain's exports rise and her imports fall. Her balance of trade and payments improves, the exchange rate of the pound sterling rises, and gold begins to return to Britain.

The reverse processes take place in the rest of the world: high prices attract commodities from Britain and restrain locally produced commodities from being exported. Exports from the rest of the world to Britain decrease, imports rise, trade and payments balance deteriorates, exchange rates in relation to the pound sterling drop and gold flows from these countries to Britain.

The operation of this spontaneous mechanism is supported by the above-mentioned policy of the central banks. In Britain interest rates went up and credit was restricted, in the other countries they went down and credit became more accessible. This had a corresponding effect on demand and prices. If we continue the comparison with the communicating vessels, we can say, for example, that in this way the communicating pipes were slightly widened and the process of levelling the liquid was accelerated.

All this sounds too good to be true. What really happened was very far from the ideal model that we have described. And this is not surprising. The model of the gold standard is based on a number of assumptions that have little or nothing in common with reality. Here are the main ones.

The theory assumes that all foreign economic operations are sensitive to prices or interest rates and change correspondingly under their influence. In fact this is by no means the case. Throughout the whole period of the gold standard Britain exported long-term capital. British money was used in other countries to build factories and railways and governments floated loans on the London market. Income from capital investment flowed into Britain just as regularly. These operations followed their own course, without, as it were, paying any attention to the movement of gold and foreign exchange rates.

No less doubtful is the assumption that the inflow of gold into a country inevitably causes an increase in money supply, and that this increase, in turn, brings about a corresponding rise in prices (and the reverse happens in the country which the gold had left). This thesis was strongly critised by Marx. If a country's economy does not feel the need for money, to force it to increase the money supply in the conditions of the gold standard is impossible. The gold imported into the country may either disappear into the reserves of private hoards, or remain dead wealth in the vaults of the central bank.

Further. Even if the money supply really does increase, this does not mean there will necessarily be a rise in prices. A most important factor for price levels is the phase of the economic cycle: prices usually rise in phases of upswing and boom. Probably the growth of the money supply helped to promote a rise in prices only to the extent and in those cases in which it coincided with these phases of the cycle.

Finally, and this is perhaps the most important point, in real life the levelling and restoration of some sort of equilibrium took place not in the form of smooth processes of adjustment, but through acute economic and financial crises, at the price of great sacrifices for the economy and the population.

The outflow of gold from the country was usually a symptom of economic crisis and accompanied it. In turn, it aggravated this crisis. When the gold left, the banks reduced their loans. The money market became "tight": it was difficult to get ready money for payments due, often for payments which decided whether or not the firm would go bankrupt. The bankruptcy of some firms brought about the bankruptcy of others, and defaults on loans led to banks going bankrupt. Credit was undermined and everybody was asking for ready money: gold or banknotes and other credit money exchangeable for it. This was called financial panic. The central bank, which in accordance with the "rules of the game" raised its interest rate and restricted credit, was actually only making the situation worse.

What happened in such a situation in the economy? Capitalist firms would sell off their inventories cheaply, reduce or cease production, sack a section of their workers and cut the wages of those who remained. Unemployment rose, the national income dropped, and the position of the workers deteriorated.

Only at this cost and in this way did an economic crisis gradually pass and the conditions arise for an upswing in the economy. Gold returned to the country and the gold reserve of the central bank was restored. In order to restore gold worth several millions or tens of millions of pounds sterling to the vaults, it was necessary to shift and put into motion the whole cumbersome economy of the country, disturbing its normal course of development.

This procedure, as one Western economist points out, reminds one of a clown who, seeing that the stool is a long way from the piano, tries to drag the piano up to the stool.

A levelling and adjustment do indeed take place, but they take place in painful convulsions with great sacrifices in a bitter struggle of everyone against everyone. If this is similar to the system of communicating vessels it is full not of water, but of a sticky liquid which clogs up the connecting tubes and passes through them in fits and starts, spurting out from time to time, etc.

One thing can be said: we should be surprised not that the gold standard in the capitalist world was so short-lived, but that it lasted as long as it did. The relative stability of money and foreign exchange that it ensured is explained not by its perfection, but by the concrete conditions in which capitalism developed in the second half of the 19th and the beginning of the 20th centuries. Among these conditions we must mention the central economic and political role of Britain which succeeded in imposing on the rest of the world the "rules of the game" of the gold standard. This system was based to a large extent on London as the commercial and financial centre of the world at that time. The central banks, led by the Bank of England, supported and protected the international gold standard, a system which gave them great power and influence.

Up to World War I no serious economist would have even dreamed of doubting the rightness of the gold standard. It seemed like the crown of creation, the supreme economic achievement, the symbol of order and bourgeois freedom. All the more surprising then are the profound and penetrating remarks of Marx on the questions contained in particular in some chapters of the uncompleted Part V in the third volume of *Capital*.

"The central bank is the pivot of the credit system. And the metal reserve, in turn, is the pivot of the bank." After noting this fact, he speaks of "the appalling manifestation of this characteristic that it possesses as the pivotal point during crises". The bourgeois economists of the various schools admit that "the greatest sacrifices of real wealth are necessary to maintain the metallic basis in a critical moment".[4]

Both in the sixties of the last century, when Marx wrote down these ideas, and in the 1890s, when they were first published by Engels, they were way ahead of their time.

Marx's ideas on gold and the gold standard are closely connected with the principles of his teaching on the nature of the capitalist mode of production and its fate. The outflow of gold, the financial crisis and their economic consequences are a manifestation of the very essence of capitalism as a system in which what is essentially social production is not subject to public control.

Unlike bourgeois economists, Marx saw capitalism not as a natural and everlasting system, but merely as a natural stage in the development of human society. It was also from this standpoint that he approached the monetary and credit system created by capitalism. The large banks and their form of "universal book-keeping and distribution of means of production" he regarded as "the most artificial and most developed product turned out by the capitalist mode of production".[5] Marx believed that this mechanism would be filled with a new content and used by socialism. It is difficult to imagine that he could have said this about gold money and the mechanism of the gold standard.

Between Two World Wars

For more than half a century economists and politicians in the West have been arguing about the consequences of restoring the gold standard in Britain in 1925. The point is that during World War I the British government, like the governments of almost all other countries, practically abolished the gold standard and tried to concentrate all the country's gold in its own hands. No longer exchangeable for gold, the pound dropped in value in relation to the gold, commodities and stable foreign currencies. City financial magnates and Conservative politicians refused to resign themselves to this: the gold pound was the symbol of the British Empire on which, according to the popular saying, the sun never set. A fitting person was found to carry out the operation of restoring it—the Chancellor of the Exchequer Winston Churchill, later Prime Minister of Great Britain. The operation was a tortuous one similar to that described above. It was necessary to bring about a drop in the level of prices in the country, a rise in the exchange rate of the pound sterling on the world market, and an improvement in the balance of payments and the gold inflow. This was achieved at the price of mass unemployment, lower wages and tightening of the belts—a policy which in recent decades has been called *deflation*.

The gold standard was restored on the basis of the pre-war

parity of the pound. This meant that the old ratio between the pound and the dollar was also introduced again. Under this ratio price levels in Britain, despite a certain drop, were significantly higher than in the USA. The exchange rate of the pound was thus raised too high. A country whose exchange rate is too high finds it difficult to export.

Let us illustrate this with the help of figures, because the question of exchange rates and foreign markets will play an important role in the rest of our argument. The parity of the pound was about 5 dollars. But in fact it was worth only 4 dollars, in the sense that a typical sample of commodities worth a certain number of pounds in Britain was expressed in the United States in a sum of dollars four times higher than that number of pounds. This difference is a measure of the extent to which the exchange rate of the pound was artificially high, the extent of the overvaluation of sterling. Let us assume that a British firm sold a commodity in the USA for 1,000 dollars. Exchanging these for its national currency, it obtained 200 pounds sterling (1,000:5). But in fact this commodity could cost more likely about £250 (1,000:4). In these conditions it was not profitable for the British firm to sell the commodity abroad.

As a general rule we can say that an artificially high exchange rate of a national currency restricts a country's exports, stimulates imports and leads to a deficit in the balance of trade and payments. A fall in the exchange rate can in certain conditions increase the competitiveness of the country's commodities and help to restore the balance of payments. Such a fall is called *devaluation.*

Let us return to Britain in the 1920s, however. Generations of economists have accused Churchill of ignorance and a fatal error. In 1975 Galbraith wrote: "The 1925 return to the gold standard was perhaps the most decisively damaging action involving money in modern times."[6]

Some are even inclined to believe that the "Great Depression" itself, the world economic crisis of 1929-1933, was a result of this mistake. Their logic is as follows. In spite of the weakening of its position in world economy after the war, Britain was still playing a major role in it as a leading importer of raw materials, financial centre, etc. The return to the gold standard on the basis of pre-war parity and the consequent heavy deflation made Britain the weak link in the economy of the West in the 1920s. In 1926 industrial production was 30 per cent below the 1913 level, and more than 12 per cent of the working population was unemployed. Britain was not successful in exporting its industrial goods and reduced its imports of raw materials. The weakness of London brought instability to the world money market as a whole. All this helped to turn the collapse of the New York stock exchange in autumn 1929 into a world crisis of unparalleled dimensions. If we like we can extend this line of reasoning further. The seizure of power by the Nazis in Germany was closely linked with the economic crisis. Thus, we arrive at a remarkable conclusion: even World War II was causally connected with Britain's return to the gold standard! It is, of course, difficult to take such hypotheses seriously. The world economic crisis was engendered by more profound factors than British financial policy, even if the latter was mistaken. But this factor could have helped to deepen the crisis. Let us recall that Marx, who always sought to get to the objective essence of economic processes, nevertheless believed, for example, that "ignorant and mistaken bank legislation ... can intensify ... money crisis".[7] He had in mind the operation in Britain of laws which strictly limited the issue of banknotes just when the economy was badly in need of money.

The universal lament of theoretical economists and financiers in the 1920s was: gold shortage! In fact the amount of gold in the state reserves was almost double that in 1913. But demand had grown even more. The inequality in the distribution of reserves was far greater: by the end of the twenties almost two-thirds of all the gold in the capitalist world was concentrated in two countries—the USA and France.[8] But the important point was this: together with pre-war capitalism the age of relative stability, international trust, and gradual change had become a thing of the past. For example, Britain's gold reserve was several times larger than before the war, but it performed its functions far less efficiently.

The post-war gold standard was merely a pale reflection of the pre-war. The circulation of gold coin had been to an insignificant extent maintained in the USA. Britain and France restored it in the form of the so-called *gold bullion standard*: the central banks exchanged banknotes for gold bars of a standard weight (about 12.5 kilograms). This meant in practice that "small fish" was removed from operations with gold, and only big capitalists could make real use of the right of exchange, particularly private banks. It is said that in France, where the gold standard survived longer than anywhere else, up to September 1936, small hoarders sometimes used to buy such a bar by clubbing together and would then have it divided at the nearest jewellers.[9] But this was no longer the normal functioning of the system, only its death throes, so to say.

The other form of economising the metal was the *gold exchange standard,* a system by which the central banks of many countries exchanged their banknotes not for gold, but for a currency which was recognised as exchangeable for gold. In order to uphold this exchangeability central banks kept considerable *foreign exchange reserves.* Whereas before the war the pound sterling was practically the only *reserve currency,* now it was strongly rivalled by the American dollar.

The point of all these types of gold standard was that gold was removed almost entirely from domestic circulation and concentrated in the hands of the state as world money, as a universal means of payment on the world market. But this system, erected with such difficulty and sacrifices, did not exist for more than a few years: the capitalist world was on the threshold of the most profound economic crisis ever known.

The break with gold in the capitalist countries took place in different ways. In Britain the gold pound collapsed in September 1931 under the blows of the international financial crisis: the capitalists of other countries refused to have confidence in it. The government announced that it regarded the measures taken as temporary ones, but, as the British writer Northcote Parkinson (author of the famous Parkinson's Law) remarked, there is nothing more permanent than something temporary. In the USA the abolition of the convertibility of the dollar into gold and the mobilisation of gold by the state was one of the first anti-depression measures of the so-called New Deal of President Roosevelt who took office at the beginning of 1933 at a time of acute crisis. France was the last of the gold standard Mohicans. Its repeal took place in a tense political situation when the bourgeoisie was fighting against the left-wing government of the popular front (led by the socialist Léon Blum), removing capital from their country's economy on a massive scale and transferring it abroad. In Germany the Nazis subjected gold and foreign exchange, as the country's whole economy, to the preparations for war. Strict limitations were imposed on all payments abroad, and foreign accounts in marks were frozen. The question of the gold standard was simply taken off the agenda.

The most important economic result of all these changes was that the fixed ratios between currencies (parities) ceased to exist. Currencies were allowed to "float", as they have in our time since 1973. As soon as a country went off the gold standard the exchange rate of its currency dropped sharply. A country often tried to let its rate drop as low as possible because this was an advantage in the competition for markets. In these conditions stable gold parities would only have been a burden. Gold was now removed from private money turnover and from the domestic sphere. The state concentrated monetary gold entirely and used it for international settlements as a final means of payment. The mechanism of the international movement of the metal also changed. Now it could move only from the vault of one central bank (or special state reserve) to the vault of another bank. Actually in the pre-war and early war years its path was always the same—over the Atlantic to the United States.

In the forties and fifties the main American repository of gold became so famous that Ian Fleming, the English writer of novels about the notorious secret agent 007 James Bond, made an attempt to rob Fort Knox the subject of one of his "thrillers" (*Goldfinger*). Goldfinger is a pathological worshipper of the yellow metal. Only the exceptional qualities and determination of James Bond prevent him from carrying out his diabolical plan, namely, with the help of the leaders of some bands of gangsters to rob the United States of gold worth 15 billion dollars (at the then official price of 35 dollars an ounce). As regards the power of gold Fleming makes Goldfinger speak most eloquently on this subject and reveals his own knowledge of it.

The concentration of world gold reserves in the USA played an important part in the foreign and external economic policy of the US government during and after the war.

Bretton Woods

In the summer of 1944 the delegates of 44 countries, future members of the United Nations, gathered in the small town of Bretton Woods (USA, New Hampshire) for an international financial conference. War was still raging in Europe and Asia and on the oceans, but victory was in sight and questions of the post-war political and economic structure of the world were arising.

The conference discussed a question previously unheard-of, the establishment of a world monetary and financial organisation and the international regulation of monetary relations. This was a new and complex matter. The conference resolved to establish such an organisation, the International Monetary Fund, and approved its Articles of Agreements, which then went for ratification by the legislative bodies of the member countries. It must be said at once that the Soviet Union did not ratify this document and did not join the Fund. Very soon after the end of the war it became obvious that the Fund would be controlled by ruling circles of the USA, whose position in relation to the USSR was becoming increasingly hostile.

Today all the industrially developed capitalist countries, with the exception of Switzerland, all the developing countries, with the exception of a few small states, and some socialist countries (Yugoslavia, Romania, Vietnam, Laos and China) are members of the International Monetary Fund. At the beginning of 1982 it had about 150 members.

Fresh in the minds of the creators of the Bretton Woods system were the hard 1930s, the years of the world economic crisis. It was taken for granted that the events in the monetary sphere (the collapse of the gold standard, the competitive devaluation of currencies, the decline of international credit, and the introduction of various restrictions on commerce and payments) had largely promoted this crisis. The fear that the post-war period held similar upheavals in store for world capitalism was widespread among economists and politicians.

The measures envisaged by the creators went in two directions. First, it was proposed to introduce strict "rules of conduct" in the monetary sphere: not to devalue without the agreement of the Fund, to abolish and not to reintroduce restrictions on payments abroad, etc. Secondly, it was planned to set up a new system of international credit which the Fund was to grant to countries with monetary problems.

In the West it is generally considered that the Bretton Woods system is largely the work of two people: the English John Maynard Keynes and the American Harry D. White. Each drafted his plan for the future organisation, and both plans were submitted to the conference. By that time Keynes was regarded as the leading light in economics, the creator of the theory of state regulation of the economy, and an authority in the sphere of economic policy and finance. Shortly before this he had been made a peer, the first aristocrat representing political economy, so to say. White was a practical economist, an Assistant Secretary of the United States Treasury. He was a "Roosevelt man", one of the group of counsellors and advisers to a president who played such an important part in the history of the United States in the 20th century.

Keynes died in 1946 greatly respected. White's fate was different. During the period of MacCarthyism and "witch hunting" he was accused of subversive activity and in 1948 summoned to answer the charges to the notorious Un-American Activities Committee. A few days later he died of a heart attack.[10]

Keynes and White shared a common aim—to strengthen the monetary and financial system and thereby consolidate world capitalism. But their plans differed greatly, because Keynes wished to restore the position of Great Britain, and White to enhance the leading role of the USA in the post-war economic structure. Keynes proposed that the Fund should grant credit generously and freely, because it was clear that Great Britain would be one of the first to ask for this credit. This was unacceptable to the USA, because it was bound to become the main creditor. Therefore White's plan envisaged far stricter terms for credit. The forces were unequal. The conference listened respectfully to Keynes, but accepted as a basis the plan submitted on behalf of the US government.

The main aim of the new monetary system was to retain the advantages of the gold standard, while getting rid of its flaws. The *advantages* were as follows: stable ratios between currencies, their mutual convertibility, and the free movement of commodities and capital. Great importance was attached to the discipline of the gold standard: if a country was "living beyond its means", it began to lose gold and was forced to take economising and restrictive measures, to pursue a policy of so-called deflation. The *flaws* were the inflexibility and arbitrariness of this system, the fact that it imposed deflation upon countries too soon and too often. And deflation meant unemployment, a slowing down of economic growth. Since the creators of the Bretton Woods system were already afraid of such things in the post-war world, this was, in their eyes, the main defect of the gold standard.

Nevertheless gold was given an important role in the new system. This was connected both with the strength of tradition and the belief in its beneficial role, and with the existence of a vast gold stock in the USA. But, as Keynes said as early as 1924, they now wanted to remove the power of the autocrat from gold and leave it in the position of a constitutional monarch. Incidentally this was done in an American, and not a British, way.

Gold was restored in its role of a measure of the international value of monetary units: each country undertook to fix the gold content of its monetary unit (currency) and maintain it. Gold was also declared the main form of international reserves, the final means of settling the balance of payments deficit.

All this in no way affected the domestic monetary functions of gold. In fact, it was taken for granted that the exchangeability of banknotes for gold would not be restored and that gold would not play the role of money inside a country. However, in most countries, in particular in the USA, by force of tradition, laws continued to operate requiring that a fixed proportion of banknotes and other types of credit money should be covered by gold. These standards were not abolished until the 1960s in the USA, and in some countries they are still in force today.

Thus, a kind of *international* gold standard was introduced. But it turned out to be not so much a gold, as a *gold dollar* standard.

The dollar now held the position of an intermediary between all the other currencies and gold. In fact, currencies were tied not to gold, but to the dollar which was considered a kind of fully-empowered representative of gold. The Articles of Agreement of the IMF read: "The par value of the currency of each member shall be expressed in terms of gold as a common denominator or in terms of the United States dollar of the weight and fineness in effect on July 1, 1944" (Article IV, Section 1).

Of real economic importance was the gold content of the dollar or, which is the same, the official price of gold in dollars, because it was at this price that it was used to settle the balance of payments deficit—was bought and sold by the central banks. From 1934 to 1971 this price remained unchanged at 35 dollars an ounce. It seemed to be a kind of natural constant in world economics, like physical constants.

Of major importance for each country was the dollar parity of its currency (or par value in terms of the Articles of Agreement), its ratio to the dollar. There were disputes around the fixing of this figure, which sometimes developed into bitter struggles. For on it depended the profitability of exporting and importing goods and the results of other foreign economic operations. After the dollar parity had been fixed, they would take a calculating machine and work out what this meant in gold. The International Monetary Fund would record both figures as being identical. The initial parity of the pound sterling was fixed at 4.03 dollars. In September 1949 Britain devalued the pound, reducing this parity to 2.80 dollars. This produced a kind of monetary earthquake, a change that had important economic and political consequences. The corresponding reduction in the gold content of the pound was of far less significance.

Having stockpiled in the armoured vaults of Fort Knox and Manhattan a vast share of the world's gold reserves, the USA could permit itself the luxury of freely exchanging dollars for gold for *foreign governments and central banks*. No American or foreign individuals, firms or banks had this right. Foreign states simply did not have enough dollars to exchange at that time. But this singled out the dollar from the rest of the company. If it wasn't gold, it was certainly gilt, so to say.

This limited exchangeability of the dollar was the thin thread which linked the whole system with gold. If a country had a currency that was convertible into dollars, the currency was by virtue of this in the final analysis and in a certain sense exchangeable for gold. In the early post-war years there was, as a rule, no convertibility into dollars, but later, as the economies of the West European countries and Japan revived and grew stronger, this convertibility was introduced. However, the dollar remained the unquestioned hub of the system.

"The dollar is as good as gold"—this financial "truth" became almost proverbial with Americans at that time. "The dollar is better than gold!" exclaimed the enthusiasts arguing as follows: the holders of dollars (accounts in US banks and American securities) receive interest on their money, whereas gold not only does not yield an income, but even requires expenditure. Today such utterances would merely produce a smile, but in the forties and fifties they had a certain meaning.

In operations between central banks gold was bought and sold mainly for dollars, and physically, as we have seen above, rarely left the USA. But it continued to play the role of the ultimate means of payment, the last resource for settling the balance of payments deficit. Dollars owned by central banks could also be converted into gold by being presented to the United States Treasury for exchange. Therefore the movement of gold between state repositories was fairly intensive.

In 1945-1949 gold flowed into the USA, as in the pre-war period: for the impoverished, war-devastated countries it was the only way of paying for American goods. By the end of 1949 the gold reserve of the USA had reached a record level—21,800 tons, which represented 70 per cent of the capitalist world's reserves. After this the flow of gold into the USA ceased for a while, and in the fifties it began to flow in the opposite direction. By the end of 1960 the USA had 15,800 tons (44 per cent), and in 1972, when the United States closed the gold window and stopped exchanging dollars for gold, only 8,600 tons (21 per cent).[11] At this level the gold reserve of the USA has remained practically frozen, like the reserves of other countries. The movement of gold between states has come to a halt.

The change in the gold reserve of the USA reflects only the final result (balance) of its movement between the USA and other countries. In fact until the end of the sixties the USA was constantly selling gold to some countries and buying from others. In the period 1951 to 1966, 13,200 tons was sold and 5,100 tons bought. The difference represented the flow of gold from the USA.[12]

The International Monetary Fund itself in accordance with its Articles of Agreement required that member countries subscribe 25 per cent of their quotas in gold. If a country did not have much gold, this contribution was established differently: 10 per cent of its gold and dollar reserves.

The formation of the post-war monetary system reflects the adjustment of state-monopoly capitalism to the changing conditions in the world. To a certain extent it corresponded to the requirements and conditions of world economic development. It ensured the conditions for a considerable expansion of world trade and other forms of economic integration of nations. From 1949 to 1970 the physical volume of foreign trade of capitalist countries quadrupled, and the growth of world trade overtook the expansion of production. This means that international division of labour increased and specialisation and co-operation of production developed. The expansion of foreign economic relations helped to raise economic efficiency.

This temporary and relative stability came to an end in the second half of the sixties.

Reserves and Debts

The following paradox exists in the world of finance: if you owe someone five dollars and cannot pay it, you are in your creditor's hands. But if you owe 5 billion dollars, your creditor is in your hands. By the end of 1967 the United States owed the rest of the world about 36 billion dollars, of which 18 billion was to governments and central banks. At the beginning of the eighties this sum rose to more than 200 billion dollars. This is both a method of US imperialist expansion and a source of instability in the monetary system.

The United States' debts to other countries are of a specific nature. They are simultaneously monetary reserves, reserves of international means of payment accumulated by other countries. Just as a coin has two sides, so, for example, any bank account looks different depending on whether you regard it from the angle of the depositor (creditor) or from that of the bank (debtor). Both these aspects constitute a single whole, but at any given moment and from any given point of view one of them can be of predominant importance. In the first two post-war decades the *reserve* aspect was at the fore and it was somehow forgotten that these were also American debts. When the accumulations reached tens of billions of dollars, however, the *debt aspect* became the main one, because the solvency of the debtor began to be doubted. This "transformation" of reserves into debts was the most important factor of the monetary crisis in the seventies.

The gold reserve of the USA was like the reserve which any bank must have in order to be able to pay cash to its clients at any moment. In normal conditions some depositors take out their money and others pay money into the bank, so that it can manage with fairly small reserves. But when confidence in the bank has been undermined and depositors become anxious about their money, a massive, panic-stricken withdrawal of money begins, known by the colourful expression of a "run on the bank".

In the first few post-war years the USA had no cause to fear such a development of events. In 1950 its gold reserve was almost seven times higher than the dollar assets of foreign powers. But in 1967 it was already only 78 per cent of the assets the owners of which could demand gold at any moment. In 1971 this percentage dropped to 22. It had reached a critical point, and at this stage the government of the USA closed the cashier's window. For a private bank this would have meant bankruptcy and selling up.

The collapse of the gold-dollar standard was preceded by some dramatic events which had important political aspects. Leading Western statesmen took part in them. A few words must be said here about France and her policies.

At the end of the fifties a period of high inflation was replaced in France by relative stability of money and prices. In ten years industrial output almost doubled. Having modernised and rationalised their enterprises, French firms launched an offensive on foreign markets, ousting their competitors. The balance of payments, which had previously had a chronic deficit, improved considerably. France began to stock foreign exchange reserves. By the end of 1957 the total amount of gold and foreign exchange (in fact dollar) reserves had dropped to a critically low level—645 million dollars. Ten years later it was 11 times higher —almost 7 billion dollars.

France began to pursue a far more independent foreign policy. It left the military organisation of NATO, stopped the colonial war in Algeria, and improved its relations with the socialist countries. France refused to follow obediently the lead of US policy.

These changes were closely associated with the name of Charles de Gaulle, the distinguished statesman who led the country from 1958 to 1969. De Gaulle was a complex, original and striking personality. Brought up on the ideas of "la grandeur de la France" and devoted to them, he was inclined to associate this with "honest" gold money. For the gold franc which had served France faithfully for more than a century had been introduced by Napoleon, the General's favourite hero. Deficits and inflation were equally firmly associated in his mind with the petty political intrigue which he hated.

De Gaulle found confirmation and support for his intuitive views, based more on emotion than reason, in the theories of influential French economists, of whom mention must be made of Charles Rist and Jacques Rueff. This school disagrees with the Keynesian school in many respects, because it sees the salvation of capitalism not in a strengthening of state regulation, but, on the contrary, in a consolidation of "free private enterprise", in preserving spontaneous market mechanisms. The gold standard, of course, is well suited to these ideas.

The views of the French "metallists" (supporters of metal money) also reflected the French passion for gold, with which we are familiar. This passion is connected with the historical peculiarities of the development of France and the social structure of French society. In the latter half of the 19th and the beginning of the 20th centuries France was the classical country of the *rentiers*, money capitalists living on incomes from loaned capital. The creditor, who expects the loan to be repaid and the payment of interest, has a vested interest in the stability of money. Gold is the most stable money. The French rentiers were ruined by the inflation which began in 1914 with only a short respite in 1927-1936, when France was on the gold standard. This increased the desire to hoard gold and the nostalgia for reliable, stable money.

In the concrete conditions of the sixties French metallism acquired a clearly expressed anti-American bias. On this account the French economists themselves leave us in no doubt. By making the dollar a reserve currency, the United States gained an advantage over other nations: it could have a payments balance deficit for many years and settle it with paper dollars which other countries were forced to accept and hold. By handing over goods and accumulating dollars other countries were increasing inflation at home. The Americans should be deprived of this possibility and made to pay their bills in gold. Then they would have to put their economy in order, whether they liked it or not, to stop rising prices and put an end to their payments balance deficit.

Rueff and those who share his views do not propose restoring the exchangeability of money for gold in internal circulation. But they believe that strengthening its role in the international monetary system could make the latter healthier. Gold should, in their opinion, be the main form of international reserves and should circulate intensively between countries.

These views obviously led to the following practical conclusion: that the US Treasury should be asked to exchange the dollars stocked by France for gold. On Rueff's recommendation the French government did in fact do this as soon as France had a "dollar surplus". On 4 February 1965 the matter was transferred to the sphere of high politics. On this day President de Gaulle made an announcement that dealt the dollar a well-calculated and painful blow.

Imagine that a banker already in difficulties with ready money is approached by one of his largest depositors who withdraws a considerable amount of his deposit. The depositor announces, moreover, that he will continue to withdraw the ready money that is on or enters his account. But even this is not all. He intends to advise his neighbours, other depositors, to hurry up and withdraw their money before it's too late. The banker is well aware that if a few more large depositors follow this advice the only thing he can do is stop paying out. This sort of thing happened in the world economy at the beginning of 1965. The role of the rich but very nervous banker was played by the government of the United States, the role of the cautious depositor by the French government, and the remaining roles by other capitalist countries that hold reserves in the form of dollar deposits. De Gaulle sharply criticised the whole monetary system of the West in terms that combined a Rueff-type economic analysis with the rhetorical style of the President: "We consider it necessary for international exchange to be established, as was the case before the great misfortunes of the world, on an indisputable monetary basis which does not bear the mark of any country in particular. What basis? To tell the truth, it is difficult to see in this respect any other criterion, standard, except gold. Yes, gold, which does not change its nature, which can be put in bars, ingots or pieces, no matter which, which has no nationality, and which has always been accepted universally as the unalterable and financial value par excellence... It is a fact that still today no currency counts except in terms of its relations, direct or indirect, real or supposed, with gold... But the supreme law, the rule of gold (and here is the place to say it) which must be put into force and honoured in international economic relations, is the obligation to ensure, from one monetary zone to another, by real returns and withdrawals of gold, the equilibrium of the balance of payments resulting from their exchanges."[13] By "great misfortunes" de Gaulle was obviously referring to the world economic crisis of 1929-1933 and World War II. Thus, he proposed restoring the gold standard of the inter-war type, the main feature of which was the decisive role of gold in international settlements.

De Gaulle's announcement was followed up by further withdrawals of gold from the USA. The dollar reserves were reduced to the level of a "working balance" necessary for current operations. Some countries followed France's example, although not with such publicity. In three years the USA lost more than 3,000 tons of gold.

The USA sought to bring pressure to bear on other countries, demanding that they refrain from exchanging their dollars for gold. From pressure the United States turned to arm-twisting and threats. To a certain extent this seemed to help: in 1969 and 1970 the US gold reserve did not drop. But this was a short respite. In 1971 a new lack of confidence in the dollar arose, and some West European countries again began to withdraw gold. This was the straw that broke the camel's back. In August 1971 the United States abolished the convertibility of the dollar into gold.

De Gaulle's successors, Pompidou, Giscard D'Estaing and

others, continued his policy, but less resolutely and more prone to compromise. The sharpness of France's "anti-dollar" attacks frequently varied under the influence of several factors, but nevertheless France's position on the main questions, including the role of gold, was invariably different from that of the USA and often opposed to it.

In the seventies the monetary problem became a constant subject of top-level negotiations. Commenting upon a speech given by the President of France and paraphrasing the well-known saying about war and generals, a French newspaper wrote that finance was too serious a thing to be trusted to the financiers. On the other hand, it is difficult for a statesman today to manage without a special knowledge of economics and finance. The figure of the professor of economics who makes a political career as a minister and statesman is becoming fairly typical. Such was the career of the British Prime Minister Harold Wilson, the West German Chancellor Ludwig Erhard, and the Prime Minister of France Raymond Barre. And Pompidou and Giscard d'Estaing were both great specialists in finance.

How much this helps a country's economy and the world capitalist economy to overcome the growing difficulties is another question. To our mind no team of professors and practical economists can rescue capitalism from the labyrinth of crises and problems which mark the 1970s and 1980s.

VIII. THE INTERNATIONAL MONETARY SYSTEM

Gold in the international monetary system of capitalism from the sixties to eighties is the subject of this chapter. To understand the place and role of gold in this system let us take a short look at its main features and the changes that have taken place in it.

People may have the impression that the fate of gold in our time is decided by conferences and committees, laws and international agreements. Some are against gold, others in favour of it, and finally a compromise is reached and some sort of agreed decision is passed. This is how things look from the outside. Never before have so many committees and councils discussed the international monetary system and the role of gold in it.

An American professor, not lacking in humour and penchant for versification, has written a variation on the English nursery rhyme of Humpty Dumpty in this connection:

Humpty Dumpty sat on a wall,
Humpty Dumpty had a great fall,
All the king's horses and all the king's men
Formed an ad hoc committee to consider the situation.

In this case Humpty Dumpty's fall is the crisis of the Bretton Woods system, which produced an attack of committee fever.[1]

Of course, a great deal does depend on the decisions of committees and even more on the policy of governments. But if we take a closer look at the question we shall see that these decisions and measures are themselves dictated by objective processes in the world economy and the financial situation. The new is born not at baize-covered committee tables, but in economic reality itself. This is obvious to any pragmatic Western financier.

The Deputy Governor of the Deutsche Bundesbank said, for example, that the retention of the present system of floating exchange rates "will be imposed on the world economy not by formulas on paper but by hard economic facts".[2]

The main turning points in the fate of gold, the abolition of the convertibility of the dollar into gold, the repeated rises in the free price of gold, the cessation of operations with it at the official price and the loss of any semblance of reality by the gold parities took place under conditions of crisis. This is also true of other changes in the system, particularly the introduction of floating exchange rates instead of fixed parities. The reform, which was agreed upon in 1976 after endless debates, merely recorded a state of affairs that had already developed. In the late seventies and early eighties new urgent problems have arisen which do not allow the committee fever to abate. Academic experts and practising financiers are producing more and more new plans and schemes. Those which reflect reality and correspond to it are being carried out to some extent or other.

Capitalism in the Last Quarter of the Century

In the last quarter of the twentieth century world capitalism has been confronted with a difficult economic and socio-political situation. The economic recession of 1974-1975 was the gravest of the whole post-war period. Even when it was gradually overcome, it left a bitter legacy which made the next upswing weaker, unbalanced and brief. The rates of economic growth in most countries had already slowed down considerably in 1979, and in 1980 the economy of the USA and a number of other countries entered a new period of recession. All in all, over the seventies growth rates in the developed capitalist countries were almost half what they were in the sixties.

All this does not mean, as Leonid Brezhnev pointed out at the 25th CPSU Congress, that there will be a kind of "automatic collapse" of capitalism. It still possesses considerable reserves and a fair capacity for adapting to new conditions. In many industries intensive scientific and technological progress continued. But on the whole the events of the seventies and the beginning of the eighties point to a weakening of the capitalist positions in the modern world. Marxists call this a deepening of the general crisis of capitalism, general crisis meaning the state in which capitalism is no longer the dominant social system in the world, but is forced to compete with socialism.

A most acute economic problem of the seventies, which has been inherited by the following decade, was the combination of high unemployment and relatively low growth rates with strong inflation. Inflation can be measured by the rise in the cost of living (see *Table 5*). As we can see in almost all these countries, both developed and developing, prices rose increasingly quickly and money depreciated more and more intensively. Although a great deal was said about inflation in the sixties also, this turned out to have been a blessed time of relative stability: in most countries prices did not rise more than three to four per cent a year. In recent years inflation has become double-

Table 5

Average Annual Rise in the Cost of Living Index
(in percentage)

Country	1961-70	1971-73	1974-78	1979	1980
USA	2.7	4.4	8.1	10.3	13.5
Japan	5.6	7.4	11.2	3.5	8.0
West Germany	2.9	6.0	4.8	4.0	5.5
Great Britain	4.0	8.6	16.1	11.8	17.9
France	3.6	6.3	10.8	9.7	13.6
Italy	3.9	7.2	16.6	12.9	21.2
Spain	4.5	9.4	18.8	13.6	15.5
India	6.3	8.6	6.9	6.0	11.5
Brazil	3.9	5.1	33.5	33.4	77.9
Mexico	2.7	7.4	20.1	15.4	26.4
Pakistan	3.5	11.1	14.2	8.6	11.7

Sources: *Mirovaya ekonomika i mezhdunarodniye otnosheniya*, No. 5, 1980, p. 154; *UN Monthly Bulletin of Statistics*, November 1980, December 1981.

digit in many countries: prices are rising more than ten per cent a year. Inflation in the United States passed this figure in 1979-1981. However, the rate at which money depreciates is particularly high in what are relatively the weaker of the developed countries, and also in many of the developing countries. For all the universal nature of inflation, there are considerable differences between countries. In particular, the comparatively low rate of price rises in West Germany is noteworthy. All these features of modern inflation are of great importance for the international monetary system.

It is typical that 1974 was a milestone both in the turning of the economic situation to recession and a sharp rise in inflation rates. At the end of 1973 an important event took place which left its mark on the whole subsequent development: the members of the Organisation of Petroleum Exporting Countries (OPEC), taking advantage of a favourable moment in the economic and political respect, nearly quadrupled the price of the oil which they sell on the world market. A new genie leapt out of the bottle—the energy crisis. This Arabic metaphor is all the more suitable, because the Arab countries of the Middle East and North Africa form a majority of the members of OPEC. In the West many people, even highly competent ones, are inclined to see a direct causal connection between OPEC's action and the economic disasters of the subsequent years. This is wrong, and sensible people do not agree with this point of view. The well-known American economist and sociologist Robert Heilbroner in his book on the fate of world capitalism in the light of the economic upheavals of the seventies compares the role of the oil shock of 1973 with that of the collapse of the stock market in Wall Street in the autumn of 1929 in the development of the world economic crisis at the beginning of the thirties. In both cases the true causes of the crisis lay deeper, and the events in question revealed the inner weaknesses of the system and accelerated the development of the crisis.[3]

Under the influence of the energy crisis and the growing struggle for markets there arose immense disproportions in international payments. Great Britain, Italy and France needed vast credits to finance their imports of oil. Although the USA is relatively less dependent on the imports or energy raw materials, it too was unable to cover them by its usual exports. The most sensational change in the world financial situation was the formation of multi-billion incomes and accumulations in the oil-producing countries. A stream of oil money gushed into the accounts of Saudi Arabia, Iran, Venezuela, Kuwait and some other countries. They could not make use of these billions in a short time to purchase useful commodities, so the money began to flow back in the opposite direction, as it were—to the world capital markets. A huge part of it settled in the form of foreign exchange (mainly dollar) reserves. From the end of 1973 to the end of 1977 the reserves of the OPEC countries rose more than five times over. The record-holder is Saudi Arabia. By the end of 1977 it ranked second in the world in reserves with 24.7 billion SDR units (29.9 billion US dollars), after West Germany which had accumulated the world's largest reserves—32.7 billion SDR units (39.7 billion dollars). Third was Japan (19.1 billion SDR units or 23.2 billion dollars).[4] True, in these figures the International Monetary Fund is valuing the gold reserve at the official price of 35 SDR units an ounce although, according to the Kingston Agreement, the Second Amendment to the Articles of Agreement on the Fund, this price was abolished. If the gold were valued at the market price the Saudi Arabian reserves would look less impressive by comparison with the reserves of the industrially developed countries which have accumulated large gold reserves.

The gold reserves of the oil-producing countries are rather small and do not reflect in the slightest their newly-acquired monetary riches. In 1977 all the OPEC countries taken together had less gold than Belgium or the Netherlands alone. This is a legacy of their former poverty and dependence. The currency reserves of many oil-producing countries continued to grow in 1978-1981 also. So far only an insignificant part of them is converted into gold.

The huge balance of payments deficits and surpluses that were formed after 1973 did not produce almost any international movement of gold. They were financed by various forms of credit, mainly thanks to the functioning of the international capital and money markets. A certain role in this process was played by the International Monetary Fund and other international financial organisations; they had to concentrate their efforts on granting funds to the developing countries which are not oil producers and which were in a particularly difficult position.

The functioning of international credit in the seventies points, on the one hand, to the efficiency and great potential of the credit system. But, on the other hand, the rising pyramid of debts, among which the proportion of short-term and dubious ones is growing, is undermining the excessively swollen credit superstructure of the world capitalist economy and is fraught with great dangers for it.

In 1979-1980 the balance of payments problem again became acute. One of the reasons for this was the new rise in oil prices. The incomes of the oil-producing countries were again rising sharply. In the meantime their recycling on the world capital market is becoming increasingly difficult for financial and political reasons. The predominant currency on this market is the dollar, and a central role is played by American banks and their European branches. But the dollar depreciated in 1977-1979, which brought losses to countries that hold their money in dollars or in Eurodollars (that is, dollar accounts in the banks of West European countries). As for the political "risks" the London *Economist* explained: "OPEC surplus countries want to diversify their assets away from the dollar; they do not want all their eggs in one basket, especially after the American response to the Iranian and Afghan crises. The freeze on Iranian assets in American-controlled banks, the partial embargo on American trade with Iran, the cut-off of American exports of grain and of high-technology items to the Soviet Union inevitably raise the question of where next. It takes little imagination to conjure up a renewed Arab-Israeli conflict which might lead to embargoes and asset freezes. Even though such fears may be unreal, they cannot help the process of recycling OPEC funds."

Here, among other variants and possibilities, the question of gold and of the investment of part of these funds in gold arises. After examining it, *The Economist* draws the following significant conclusion: "If one other prediction can be chanced it is that the monetary use of gold will be greater in the next 10 years than it has been in the past 10..."[5] We shall return later to the main aspects of this problem, but here we must touch briefly upon another question closely connected with gold and its monetary role—the system of foreign exchange rates.

A most important fact in the development of the international monetary system in the seventies was the forced transition from fixed currency parities to floating exchange rates—permanent and considerable fluctuations of the exchange ratios between currencies that develop on the market.

For centuries it was thought that in the sphere of gold and currencies stability must reign. Today abnormality has become the norm, and instability the only stable phenomenon.

Floating Currencies

With the gold standard, as we know, foreign exchange rates fluctuated within narrow limits around parities (par values) determined by the gold content of each currency. The stability of the parities and narrow margins of the fluctuations in exchange rates was also an important principle of the Bretton Woods system, although here this was achieved by different methods. True, in this system it often happened that a country did not fix the par value or, after formally fixing it and registering it in the International Monetary Fund, did not in fact observe this par value and did not support it on the market. If in such cases the exchange rate was not fixed by the non-market method with the help of exchange control, it could fluctuate on the market. However, first, such a regime was considered temporary and the establishment of a fixed parity remained the aim; secondly, this concerned the currency in question, but by no means the whole system. The currency in question fluctuated in relation to a kind of stability axis, but the axis itself was immobile. The gold-linked dollar acted as such an axis. In the present system there is no stability axis at all. Gold as a measure of the official value of currencies has been excluded entirely from the system, and no currency is exchanged for it according to a fixed parity. The dollar is, so to say, floating in the common pool, like all the others, although its massive figure stands out among the rest of the swimmers.

An exchange rate is a value ratio between two monetary units, it connects two currencies. If one of them drops, the other rises, and vice versa. For the sake of clarity the following analogy can be drawn. If the price index rises, the purchasing power of money falls. And the reverse, if the index drops (which, incidentally, never happens nowadays), the purchasing power of money rises. In the same way, if the dollar falls in relation to the West German mark, then the mark rises in relation to the dollar.

However, before 1971-1973 the exchange rate between a given currency and the dollar was always seen from one angle only, how this currency was related to the dollar. Again the dollar was the axis, the point of reference for the whole system. Today things are different. At the end of 1977 and throughout 1978 the steady fall in the dollar rate caused great anxiety in the West. Precisely this aspect was at the centre of attention, although one could with equal grounds have spoken of the rise in the rate of other currencies, particularly the mark, yen and Swiss franc. In the same way in 1980-1981 the opposite trend became pronounced—the rise in the exchange rate of the dollar.

The floating of currencies is closely linked with the process of removing gold from the international monetary system. Here is another paradox: the crisis of the seventies was a crisis of reserve currencies, in fact, the crisis of the dollar. But the "victim" was gold: at least, it was officially pushed into the background of the system. Throughout the spring and summer of 1971 tension rose on the exchange markets. Lack of confidence in the dollar grew to such an extent, that private holders of dollars (banks, firms and individual capitalists) sought en masse to get rid of American currency and exchanged it for more stable currencies. The dollar rate dropped to the lowest level at which the central banks, in accordance with international agreements, had to support it. This meant buying up more and more billions of dollars on the market and increasing their already huge reserves.

According to the rules of the game the dollar should have been devalued, in order that the conversion of money from dollars into other currencies became more expensive. But the USA could not do this without reducing the gold content of its currency, which had been unchanged since 1934 and had become a kind of political fetish. Another solution would have been for other countries to raise the parities of their currencies in relation to the dollar. But they did not wish to do so: as we have seen above, raising the exchange rate weakens the competitiveness of a country's commodities and hampers exports. The time has passed when the United States could impose such solutions on its partners. What is more, the partners lost their patience and hinted unambiguously that they would demand gold in exchange for their dollar accumulations.

The advisers of Nixon, who was President at that time, suggested closing the gold window (abolishing the convertibility of the dollar into gold), introducing a special additional duty on the import of foreign commodities into the USA, and also taking certain internal measures against inflation. On the evening of Sunday, 15 August 1971, in an 18-minute speech Nixon announced this "new economic policy", which was a great blow to the United States' partners. As an American weekly described it, the news got the British Prime Minister Heath out of bed in his country residence, and "by Monday morning Cabinet ministers all over Europe were flying into their capitals, dazed and tanned, from interrupted vacations in the south of France and on the grouse-filled moors of Scotland".[6]

But the ministers could not do much, at least by way of emergency measures. It had become quite pointless to buy up on the market and accumulate the "new", i.e., unconvertible into gold, dollars, and central banks stopped doing this and supporting the dollar rate. For a few weeks the exchange rates of most industrial countries rose in relation to the dollar. The system of fixed parities was disrupted, and currencies began to float, with fluctuations in exchange rates reaching considerable dimensions.

A tense situation arose, fraught with future complications. In the latter months of 1971 the flurry of conferences and negotiations reached its height. The threat of chaos made it necessary to seek a compromise. The Americans conducted these negotiations from the viewpoint: we're all in the same boat, so we mustn't rock it too much. The Europeans replied that it was Nixon who had rocked the boat, and it had already shipped a fair amount of water. The representative of West Germany compared sitting in the same "monetary boat" as the United States to "being in the same boat with an elephant".[7]

The Bretton Woods system was splitting at all the seams, and the connection of currencies with gold had been lost. But people's thinking was still dominated by familiar forms. A compromise was therefore sought in the establishing of new gold parities and restoring more or less fixed ratios between currencies. Politicians' minds were not yet ready for universal floating, although many economists had long been urging it. The change to free floating of currencies was supported particularly vehemently by the monetarists, the representatives of the school whose leader is generally considered to be the American professor, Nobel prize winner Milton Friedman. The main idea of Friedman and many of his supporters is interpreted as follows by the Soviet economist D. V. Smyslov in his book: "Thus it follows that, whereas during the period of the gold standard the free movement of gold had been the automatic regulator of international payments, now this function was to be performed by the 'floating' rates of currencies."[8] But up to the beginning of the seventies, when objective conditions in the world changed, these ideas remained largely confined to the academic sphere. Now Friedman's hour had come—in the sense not that concrete measures were taken according to his projects, but that events had taken such a turn that the financial policy-makers of the Western countries had no other choice. This, incidentally, took another two years of crises and panic. At the end of 1971 they still hoped to manage by making some minor adjustments to the Bretton Woods system.

A compromise was first reached in the bilateral talks of the presidents of the USA and France, at which the latter was to a certain extent speaking for the whole of Western Europe. Nixon and Pompidou "after much private discussion of monetary matters—a subject that fascinated Pompidou but, it is said, made Nixon's eyes glaze over—agreed that the dollar would be devalued by raising the price of gold from $35 to $38 per ounce".[9] In December 1971 this compromise was made the subject of a multilateral agreement in Washington.

This price did not and could not have any real economic significance. On the free market gold cost about 44 dollars per ounce at the end of 1971, and 65 dollars at the end of 1972. All operations were carried out at these prices, and no one sold gold at 38 dollars. Thus, that was not the main point. Politically the devaluation of the dollar was a capitulation by the government of the USA. American presidents, secretaries of the Treasury and other high-ranking officials had declared countless times that the gold parity of the dollar was sacred and that the dollar would never be devalued, implying that the other currencies were no match for it. Now the French had some cause for malicious rejoicing: the dollar had proved susceptible to the same defects as all other currencies.

But the United States was sufficiently strong and influential to extract the utmost benefit from its concession. Great Britain and France retained the former formal gold content of their currencies, Japan and West Germany raised it slightly. As a result the parities of all these and many other currencies to the dollar were raised to different degrees. This was precisely what the USA wanted: with a cheaper dollar it was easier to compete on world markets.

Nixon hastened to declare these decisions the most important agreement on the monetary problem in world history. In fact, the American President's declaration turned out to be one of the most shortsighted in world history. The system established in December 1971 lasted only 14 months. The US balance of payments did not improve and the lack of confidence in the dollar did not abate. The first step is the hardest: the dollar that had lost its "innocence" went through another devaluation in February 1973 without any pangs of conscience. The price of gold was raised to 42.22 dollars. In the meantime it reached 112 dollars on the free market at the end of 1973, and 187 dollars at the end of 1974. One day the quoting almost reached 200.

This was, as commentators with a rhetorical turn of phrase like to say, the end of an era. No one tried any longer to establish new fixed parities. The era of floating currencies began. The diagram of exchange rates after March 1973 looks like a fan. The rate of the Swiss franc, calculated on average in relation to all other currencies, shot up and in 1979 was almost twice as high as the level for the first quarter of 1973. The reason for this remarkable "stability" of little Switzerland's currency lay in its lowest rate of inflation in the Western world and the constant flow of capital from abroad, which produced a large and stable demand for its currency. Somewhat less effectively, but with iron consistency, rose the rate of the West German mark, which in 1979 cost about 45 per cent more than before it went floating. The third country with a "strong currency" is Japan. West Germany and Japan hold second and third place respectively in their gross national product in the capitalist world, and in commodity exports West Germany has even surpassed the United States. Both countries, although suffering from inflation, have managed to keep it within tolerable limits. They hold firm positions on the world markets and are coping satisfactorily with huge payments for imported oil. Incidentally, in 1978-1979 the rate of the yen dropped markedly, reflecting a deterioration in Japan's economic position. At the other end of the scale was Italy, whose currency (again on average in relation to other currencies) has lost more than 40 per cent of its value. The currency of Great Britain was also in a very bad state. Before 1977 the exchange rate of the pound sterling plummeted down. Subsequently it recovered somewhat as Great Britain's balance of payments was helped by North Sea oil. In 1979 the exchange rate of the pound was on average about 25 per cent lower than in the first quarter of 1973.

At the centre of the diagram, fluctuating around the index of 100, are the exchange rates of the American and Canadian dollar and the French franc. The fate of the dollar was particularly changeable: after a sharp drop in 1973, it recovered its former level and held it until the middle of 1977. Throughout the next year there was a sharp fall in the dollar which had important economic and political consequences; one of them was the sharp rise in the dollar price of gold, of which we shall speak later. As a result of these fluctuations the average exchange rate of the dollar in 1979 was about 10 per cent lower than its initial level.

In 1980-1981 the situation changed again. The exchange rate of the dollar began to rise sharply owing to many factors. One of them was an extremely high level of interest rates in the USA, which led to a great inflow of money capital from Western Europe and the consequent demand for dollars. In its turn, the rise in the exchange rate of the dollar played a certain part in lowering the dollar price of gold at the time.

In modern conditions the exchange rate and its movement is an important indication of the economic and financial status of a country. But the constant sharp fluctuations in exchange rates are introducing yet another element of instability into the world capitalist economy. It turns out that capitalist countries cannot allow full spontaneity in this sphere and are compelled to regulate the movement of exchange rates by various methods. They use the traditional form of *currency intervention,* i.e., the buying or selling of a currency on the market by central banks with the aim of influencing its exchange rate. Another method of somehow stabilising exchange rates, closely related to this, is to attach the currency to a leader currency and float with it.

According to statistics of the International Monetary Fund, in mid-1979 of the 137 member countries only 41 were allowing their currencies to float freely or had not chosen a regime. The currencies of 56 countries were attached to leader currencies, 36 of them to the dollar. These were almost exclusively the developing countries. The exchange rates of the currencies of 35 countries were determined on the basis of composite currency baskets, particularly the SDR units. Finally, the eight Common Market countries joined the European Monetary System and supported fixed ratios between their currencies, thereby floating together in the stormy currency sea.[10] All this is like a complex mosaic, and we do not propose to analyse it further here. We would merely note a matter which is directly related to the subject of gold.

In an abstract examination of the question economists concluded that floating currencies greatly reduce the need for gold and foreign exchange reserves, and under "clean floating", i.e., without any intervention from governments and central banks, do away with it altogether, as it were. This means the following: when the balance of payments gets worse the exchange rate of the currency drops so low that factors arise which spontaneously restore the equilibrium and no expenditure of reserves is required. In fact, it turned out that neither a country itself, nor its partners, can permit the exchange rate to drop this low. In order to regulate the balance of payments and support the exchange rate by market methods, reserves are essential. In the seventies the fluctuations in the balance of payments reached huge dimensions. Therefore the problem of reserves—their amount and structure—was not alleviated at all. This, as we shall see later, can give gold—the most reliable form of reserves—a second wind.

Paper Gold

In recent years in the world financial and monetary system there have appeared many innovations, unusual and at first glance strange. Perhaps the most important of them is "paper gold"—international credit money.

Einzig remarks that if gold had not existed, it ought to have been invented, such is its utility for the world economy.[11] The irony is that "artificial" gold was invented precisely in order to drive out and eventually replace in inter-state circulation the gold defended by Einzig. We are not speaking of a synthetic material with gold-like properties, of course, but of a special form of credit money which is based on the collective credit of the governments of many countries.

International credit money was first introduced in 1969 as a result of protracted and difficult negotiations. It became called Special Drawing Rights and is usually referred to by the initials SDR.

The questions associated with SDR and their role in the international monetary system are by no means simple ones. An American expert who took part for many years in the international negotiations on the reform of this system writes that the technical aspects are so complicated that one could not expect most of the ministers of finance officially representing their countries in the negotiations to understand them.[12]

I do not know about the ministers, but for the general public and even for professional economists who do not make a special study of the international monetary problem, SDR are certainly still something of an enigma. The International Monetary Fund publishes a voluminous monthly bulletin entitled *International Financial Statistics*. Formerly it began with a fairly simple table of par values expressed in gold and dollars. Now we learn something not very meaningful from this introductory table: how many US dollars an SDR unit is worth and how many SDR units a dollar is worth. What is more, these amounts change every day: for example, on 1 March 1977 the dollar quotation was 0.864165 SDR units, and on 31 March, 0.862691 units. The dollar had fallen slightly over the month. Another table shows the daily fluctuating exchange rate of each of the leading currencies for the dollar.

Formerly there was a series of tables reflecting each country's operations with the Fund and its foreign exchange reserves. All these amounts were expressed in dollars, but, if necessary, one could easily calculate them in gold or in any hard currency. This was also familiar and easy to understand. Today, however, we learn that at the end of 1976 the total amount of West German reserves was 29.9 billion SDR units, including 4.1 billion units in gold, 22.0 billion units in foreign currency, 1.7 billion in SDR themselves and a reserve position in the Fund (the right for its credit) of 2.1 billion.[13] The matter is even more complicated by the fact that the estimate of gold in SDR units is completely conventional: it does not in any way reflect the real market price of gold, but merely the fact that at one time the gold content of the unit was regarded as equal to the gold content of the "old" (before the 1971 devaluation) dollar and was 0.888671 grammes. Today, however, the SDR unit has no gold content and is not, of course, exchangeable for gold. As for the dollar price of gold, it is quoted most significantly in the section of commodity prices, where Gold comes in alphabetical order between Fishmeal and Groundnuts. In this way the authors of the bulletin wish to show that gold has been demonetised and reduced to the level of an ordinary commodity.

Where do SDR come from and what functions do they perform? Why are they called paper gold in the journalists' jargon?

The idea of creating this type of credit money had been in the air for a long time. It was a question of introducing into inter-state circulation the sort of money the supply of which could be supplemented consciously as necessary. In this respect it would be better than gold, the flow of which into central reserves depends on various chance factors, such as the discovery of new deposits and the demand for gold for non-monetary purposes. And it would also be better than national currencies (dollars) the use of which as international money had led to the crisis. This is how it looked on paper, in the schemes of economists.

In fact everything turned out to be more modest, but the supporters of introducing SDR are not downhearted, believing that this is only the beginning. Just as a country's central bank issues (puts into circulation) paper money, in more or less the same way the International Monetary Fund "created" and distributed, in 1970-1972, 9.3 billion SDR units. At that time the gold content of the unit was equal to that of the dollar. This money, which had appeared out of thin air, was included by each country in its reserves. Naturally, the capitalist world did not become any the richer for it, because wealth lies in real values, and this was a paper value. But for each country taken separately this addition was a real one: it increased its liquidity, its ability to settle its international obligations.

In what way do SDR resemble gold? First, in using them to

settle its balance of payments deficit, a country makes a final payment, as if it had paid in gold. It is not credit which must eventually be paid off. Secondly, SDR reserves, unlike dollar reserves, are not at the same time someone's debts. Their fate does not depend on the position of the debtor, for example, US banks. Thirdly, once SDR have been put into the circulation of central banks they cannot leave it, but can only move from one country to another or between the member countries of the International Monetary Fund.

But, of course, they are not gold. They have no intrinsic value and, like all paper-credit money, can circulate only on the basis of confidence in them and in the body which issues them. The advantage of SDR is simply that they are based on collective credit. The banknotes of the state central bank, when they were first issued in the 18th and 19th centuries, were more reliable than the banknotes of private banks, but this did not mean they stopped being paper. SDR are more reliable than national banknotes and other forms of paper-credit money, but their nature is the same. Incidentally, SDR exist only as entries in special accounts of countries in the Fund, and not in the form of banknotes.

The SDR system bears the mark of compromise. The USA was seeking to create international money that would have extensive money functions and could seriously oust gold as a means of reserves and settlements. France, traditionally attached to gold, did not want this and tried to reduce the matter to the creation of another form of international credit, like the other credit resources of the International Monetary Fund. And SDR proved to be such a monetary-credit hybrid. As one expert put it, the SDR is like a zebra: you can say it is a black animal with white stripes, or a white one with black stripes. The Special Drawing Rights created in this way played a very modest role in the turbulent period of the mid-seventies, when balances of payments experienced great changes owing to the rise in the price of oil and for other reasons. In these conditions the modest sums of the SDR whose use by deficit countries was restricted, since they were not completely money but also credit sums, could not play any real role in the balancing of payments. For six years there was no new issue of SDR.

But in 1979 the issue of SDR was renewed, and in 1979-1981

another 12 billion units were distributed among the member countries in proportion to their quotas. As with the first distribution, the bulk of the new SDR went to the industrially developed countries. Moreover, the *monetary properties* of the SDR were extended, i.e., opportunities were increased for their use by holders for buying currencies, settling obligations, etc. All this is seen by SDR supporters, whose citadel is naturally the International Monetary Fund, as steps towards the realisation of an aim now recorded in the Fund's Articles of Agreement, namely, the turning of SDR into the main form of international reserves.

For a long time now a plan has been discussed which, if implemented, would lead to a considerable increase in the supply of SDR in international circulation and, consequently, their proportion in international reserves. This plan has been called the creation of a *substitution account*. In fact, it is a revival of an idea advocated by Robert Triffin, a leading supporter of the establishment of a world bank and real international credit money, as early as the beginning of the sixties. The central banks, which have accumulated large stocks of dollars, would exchange them for SDR which the International Monetary Fund issues. Today it is being suggested that the Fund should issue special bonds in SDR for this purpose. The dollars received by the Fund would be invested in US state securities. When the amount of dollars that could be presented for such an exchange is discussed, various figures are suggested—from 10 billion to 50 billion SDR units (at the beginning of 1980 this amounted to almost 65 billion dollars). Quoting these figures *The Economist* remarks that "even the upper figure looks pretty small beer compared with the size of the problem",[14] i.e., compared with the mass of dollars accumulated outside the USA. In keeping with the spirit of the times there is now an attempt to find a place for the real gold too in this plan. It is assumed that the gold reserve of the International Monetary Fund will be used as a guarantee fund, particularly for covering the Fund's losses in the event of a drop in the exchange rate of the dollar.

Could the International Monetary Fund eventually turn into a world central bank like the Federal Reserve System of the USA, and could SDR become the only or at least the main form of international payment means? Marxists reply to this question as follows: it is impossible as long as capitalism remains capitalism. Such a degree of organisation on a world scale is contrary to the nature of capitalism, incompatible with it. The idea of world monetary super-regulation is akin to the once fashionable concept of ultra-imperialism, which was criticised by Lenin. He wrote, inter alia: "There is no doubt that the trend of development is *towards* a single world trust absorbing all enterprises without exception and all states without exception. But this development proceeds in such circumstances, at such a pace, through such contradictions, conflicts and upheavals—not only economic but political, national, etc.—that inevitably imperialism will burst and capitalism will be transformed into its opposite *long before* one world trust materialises, before the 'ultra-imperialist', world-wide amalgamation of national finance capitals takes place."[15]

Just as under capitalism there can be no question of a single world trust regulating all production, all trade and all consumption, so there can be no question either of a world superbank regulating money and credit on a global scale.

However, *movement* in this direction is possible; what is more it is logical and even inevitable. Hence the creation and evolution of the SDR. The role of the SDR as a *unit of account* in international monetary relations is most appreciable. This is understandable: when the gold parities of currencies were practically, and then juridically, abolished and currencies were floated the question arose of how to account monetary sums if they are expressed in different national currencies. What common denominator is possible?

These are the functions that the SDR unit has taken upon itself. But it is a very peculiar common denominator, because it is floating together with the currencies. The value of the SDR is not a stable one either in gold or in any sample of commodities. To the extent to which currencies depreciate under the influence of inflation, the SDR unit depreciates with them. Since the SDR unit was introduced more than ten years ago, its real purchasing power has dropped about three times. It merely averages out this depreciation avoiding extreme cases.

The value of the SDR unit (its ratio with currencies) is determined on the basis of the actual exchange rates—the so-called "standard basket" of the sixteen main currencies. Since 1981 the number of currencies comprising the basket has been reduced to five. But if this is the case, how do currencies measure their value in SDR? This is a vicious circle. In order to overcome it, a kind of point of reference was accepted, the initial value of the SDR unit in dollars and other currencies was determined. It was 1.20635 dollars and was obtained by dividing the gold content of the SDR by the gold content of the twice (in 1971 and 1973) devalued dollar. As we can see, here too gold was involved. The ratio of the SDR to all the other currencies was determined by proceeding from this figure and the exchange rate of the given currency in relation to the dollar. After this the SDR unit expresses, in fact, not the value of currencies, but the deviation of the exchange rate of each currency in relation to all the others from this initial level.

Measurement in SDR units softens, smoothes fluctuations. For example, from August to December 1977 the exchange rate of the dollar for the West German mark dropped by 9 per cent and for the Japanese yen by 10 per cent, whereas its ratio to the SDR dropped only by 4 per cent.[16]

Big fluctuations in exchange rates make creditors, investors and other receivers of future payments seek a guarantee against losses. To a certain extent the SDR provides this: for example, a loan in pounds sterling is granted on condition that a sum of pounds equivalent to a constant amount of SDR is returned.

But, as we have seen, the SDR is no guarantee against the main danger, inflation. The OPEC countries agreed long ago to quote oil prices not in dollars, but in SDR units. However, for a number of reasons this decision was not put into force. One reason is that it is not very effective: the exchange rate of the dollar for the SDR unit drops very little or even actually rises as in 1980-1981, although dollar prices of commodities may rise sharply, which means a real depreciation of the dollar.

Naturally, the International Monetary Fund calculates all monetary sums in SDR. But in recent years many other international financial institutions have begun to use the SDR as a unit of account. They include the African Development Bank, the Arab Monetary Fund, the Islamic Development Bank and others. On the international capital market bonds issued are expressed in SDR, and the large commercial banks accept deposits expressed in them.[17] Let us stress once again that no real payments are usually made in SDR as such (this is not permitted by their status), only monetary sums are expressed in them.

The use of the SDR as the unit of account satisfactorily averages out currency *exchange rates* and serves as a means of insurance against rate losses. In order to retain *real purchasing power* of a sum of money, it is necessary to use *indexation,* i.e., to recalculate the sum of money in accordance with changes in the price index. This indexation is increasingly being applied in capitalist countries. For example, parties to a contract stipulate that the contractual price of a commodity is 1,000 dollars, but will be reviewed at the moment of payment if the wholesale price index has changed more than 10 per cent by then. If the index rises by 10 per cent, the price will be 1,100 dollars, if it rises by 20 per cent, the price will be 1,200, and so on. However, indexation, or the use of index clauses, has major shortcomings and can be a technically complex business.

The question naturally arises as to whether gold would not be suitable for these purposes. Although it is now included in the list of commodities between Fishmeal and Groundnuts, the price of gold differs economically, historically, politically and psychologically from the prices of these prosaic articles. True, the price of gold fluctuates sharply, but it tends to reflect in the long run the inflationary depreciation of the dollar and other paper money. There are signs of a revival of the gold clause—the contractual condition according to which sums of money are revised depending on the price of gold.

In general gold is a dangerous competitor to SDR. Whatever the international guarantees of the SDR, the lustre of the precious metal provides better ones. Even quantitatively it surpasses this artificial creation of the financial alchemists. The total gold reserves valued according to the market price are many dozens of times larger than SDR reserves.

However, the SDR has yet another competitor, not so illustrious, but of similar lineage: in 1979 the member countries of the European Economic Community created the European Monetary System with its own composite monetary unit called the European Currency Unit (ECU). The abbreviation is reminiscent of the mediaeval French gold and silver coin.

It is also a type of paper gold but, unlike the global SDR, what we have here is regional international credit money. The sphere of its use is the settlements between member countries at the level of their central banks. It is typical that the ECU is more closely linked with gold than is the SDR. This money is issued by the system's central organisation, the European Monetary Co-operation Fund, to each country in return for a contribution (technically this is done by a swap renewed every three months) of 20 per cent of its dollar and *gold* reserves. The gold is valued at the market price. Therefore the supply of ECU issued against gold is far bigger than the amount of them issued against dollars. Thus a currency has emerged with an extremely high gold backing. The ECU is the main unit of account within the Common Market.

The European Monetary System is still in the process of formation. In the next few years its features will take more definite shape. For the time being it is obvious only that a monetary bloc that is to a considerable extent opposed to the dollar is being formed in Western Europe and that in this bloc gold will evidently play a more significant role than in the global system of the International Monetary Fund, where the USA holds a dominant position.

Reform and After

No sooner had it overcome the difficulties of the early postwar years than the Bretton Woods system began to experience a growing tension connected, primarily, with the relative weakening of the US position in the world economy. Already at the end of the fifties there was talk in academic circles of the need for reforming the Bretton Woods system. Projects for such a reform soon began to mushroom. In the sixties a mass of literature appeared on this problem. The laurels of Keynes and White—the creators of Bretton Woods—did not let many sleep peacefully. By the end of the decade it looked as though any economist and politician worth his salt felt obliged to produce some kind of "plan" or at least additions and amendments to someone else's plan.

As a rule, the Anglo-American plans proceeded from a reduction of the role of gold and the introduction of some form of international reserves. Extreme versions proposed the establishment of a world super-bank and the complete replacement of gold and national currencies in the sphere of settlements between countries by international money which this super-bank would issue systematically. True to tradition the French proposed the reverse, i.e., enhancing the role of gold and raising its official price sharply, that is, devaluing the dollar and other currencies with it. There were quite a few exceptions, however. In Britain Einzig campaigned for gold and explained the "anti-gold" position of US and British policy-makers as a sour grapes complex: like the fox in the fable they proposed renouncing something of which they had increasingly little.

At the same time it was gradually becoming more and more clear that a return to the gold standard would come up against not only the stubbornness of the Anglo-Saxons, but also the objective difficulties, the existing realities of the development of the capitalist economy. Attention was focussed increasingly on the creation of new reserve mediums. The matter shifted from the sphere of projects and discussions to the sphere of negotiations and financial diplomacy, as a result of which the SDR arose.

But the introduction of the SDR coincided with a series of severe crises in the monetary system, which took place in 1971-1973 and undermined the foundation of the Bretton Woods system improved by Special Drawing Rights. Floating exchange rates became a fact. The question of a radical reform of the international monetary system arose once more.

In the autumn of 1973, after certain principles of reform had been agreed in endlessly protracted negotiations, the oil crisis broke out. Events again shuffled the cards of the monetary reformers. After the session of the International Monetary Fund in autumn 1974, at which it was intended to approve the reform, it turned out that the efforts to reconstruct the international monetary system had three-quarters failed. The remainder boiled down to juridical recognition of a situation which had developed spontaneously in the course of severe crises and clashes of interest. At the end of 1974 the American economist J. Marcus Fleming wrote: "One of the most ambitious attempts ever made to achieve a synchronised and many-sided reconstruction of the international monetary system by an orderly process of multilateral consultation and agreement is now grinding to a halt."[18]

True enough, the mountain had produced a molehill. The points gradually worked out in 1975 and accepted formally at the beginning of 1976 at the conference in Kingston did not make any significant changes in the system. Incorporated in the form of amendments to the Articles of Agreement of the International Monetary Fund, this mini-reform came into force in April 1978, when the new Articles of Agreement were ratified by a sufficient number of members. The content and meaning of the reform has been discussed in many works and we do not propose to examine it in detail here. We would refer the reader, for example, to the definitive work of the Soviet economist D. V. Smyslov.[19]

Let us note merely that floating exchange rates received official sanction in the Articles of Agreement of the International Monetary Fund. A return to fixed parities remains just a faint possibility. The real situation in this sphere has been described above.

Although this is not recorded in any clauses of the Articles, the status of the dollar as a reserve currency in fact remains unchanged. Whereas gold can no longer serve as the main medium of international reserves and settlements, and "paper gold", the SDR, cannot yet play such a role (and it is not known whether it ever will), the dollar necessarily retains the central position.

Naturally, within the framework of our subject the question of the role of gold in the reformed system is of special interest. The essence of the Jamaica decisions which, like the agreement on the SDR, arose as the result of a compromise (first and foremost, between the USA and France) is as follows.

1. Gold shall be removed from the system as the basis for assessing the value of national currencies and of the SDR unit. If a country wishes to fix the par value of its currency it can do this by attaching it to another currency, the SDR unit or another unit of account, but certainly not by fixing its gold content.

2. Gold shall be removed from the operations of the International Monetary Fund. The former requirement of the Articles of Agreement that member countries should pay one-quarter of their quota in gold shall be annulled (now a quarter of the increment in quotas is paid in SDR.)

3. The International Monetary Fund shall pursue a policy aimed at ensuring that the free market of gold is not subjected to state regulation and that no fixed price is established for it

(as in the age of the Gold Pool). In other words, the Fund is to help to turn gold into an ordinary commodity.

4. The International Monetary Fund shall begin immediately to liquidate its own gold reserve formed from the contributions of member countries. In 1976 the question of one-third of this reserve was solved. Half of this third (i. e., one-sixth) was to be returned to the member countries who contributed the gold in exchange for their national currencies at the official rate of 35 SDR units per ounce. The other half was to be sold on the free market. Under pressure from the developing countries, whose role in the IMF has grown somewhat, the income from this operation (the difference between the market and old official price) was to go to the poorest developing countries. The fate of the remaining two-thirds of the Fund's reserve was to be decided later (more about this below).

These measures mean the formal demonetisation of gold, the abolition of its official monetary functions in the international monetary system established by the agreement. The IMF Committee which drafted the reform stated: "The objective should be an enhancement in the role of the SDR as the central asset in the international monetary system and, consequently, a reduction of the role of gold."[20] As already mentioned, the aim of turning the SDR into the main form of reserves was written into the revised Articles of Agreement of the IMF.

However, no sooner had these decisions become known than a debate began on whether the role of gold was really being reduced or enhanced. The fact is that the annulment of the official price of gold fully legalises the free price. The real purchasing or paying power of gold reserves rises. Now (since 1978) the central banks have the opportunity of buying and selling gold among themselves at the high market price. Thus, the reserve function of gold has not been reduced, but rather enhanced. Since it is obvious that no international agreement can regulate this function, the IMF Articles of Agreement pass the question over in silence: nothing is said about the role of gold as a means of international reserves, nor is there any reference to the "annulment" of this role.

Each of the traditionally opposing parties in the gold dispute could and did interpret its new status as the implementation of their principles and objectives. The American demonetisers maintained that the idea of a monetary system divorced from the spontaneous gold factor and susceptible to flexible international regulation was being put into practice. On the other hand, the French "metallists" emphasised the legalisation of the high market price of gold and the increase in its real reserve and payment possibilities.

The situation is such that some are speaking of ousting and others of the victory of gold. An article on the gold decisions published in the influential French newspaper *Le Monde* bore the headline: "Let us demonetise gold, long live gold!" (a variation on the well-known saying "The King is dead, long live the King").[21]

Gold is being driven out of the door (through the main entrance of the International Monetary Fund in Washington) and returning through the window of the free market. The yellow devil is persistent and stubborn.

The new role of gold is seen in real operations, in which it functions at the free market price as a means of international payment. Even before the Jamaica conference a credit transaction between the central banks of Italy and West Germany attracted general attention: the former used part of its considerable gold reserve as security of a large international loan in West German marks. The use of gold as ECU collateral within the framework of the European Monetary System is far more significant.

With the abolition of official gold parities and, consequently, of official gold prices, governments and central banks were faced with the practical question of how to value their gold reserves. This seemingly straightforward matter of book-keeping acquired political overtones. France had been the first to introduce as early as 1975 valuation of gold at the market price (the average for each quarter preceding the half-year for which a firm estimate is being made). As a result this item in the balance of the Bank of France soared several times and then jumped up and down every half-year. Following this many countries revalued their reserves, making use of various formulae based on the market price of the metal. The United States has not followed this example so far and values Treasury gold at an artificial price (the last official one) of 42.22 dollars an ounce, which was annulled in the International Monetary Fund with its consent. But this is regarded in the world as a stubborn anachronism.

Like the reform itself, the subsequent trends in the sphere of gold can permit of a dual interpretation. On the one hand, gold is no longer the basis of the international monetary system and appears as a *commodity,* the price of which changes on the market like the prices of other commodities. On the other hand, the potential and actual economic role of gold reserves has grown thanks to repeated rises in the price of gold, and in the function of international reserves it is demonstrating its advantages over inflationary national currencies and international credit money. This duality is important for the future of gold.

Gold from the Auction

An auction is a well-known method of selling a commodity in which the commodity goes to the highest bidder. The possessions of bankrupt debtors are usually auctioned. Auctions of works of art are also common. But the auctioning of gold is an innovation of the seventies, linked with the changes that have taken place in the functioning of gold in the capitalist economic system.

As already mentioned above, the participation of the central banks of capitalist countries in the international movement of gold was frozen first by the agreement to abolish the Gold Pool (March 1968), and then by a unilateral decision of the USA to abolish the convertibility of the dollar into gold for foreign central banks (August 1971). However, for various reasons the main capitalist countries already in 1973 yielded to the need to change this situation, which was freezing central gold reserves and reducing their real economic significance. The United States proceeded from the fact that within the framework of the policy of demonetising gold which it pursued it was in its own interests to stimulate the transfer of gold from central reserves to the sphere of the private market, where gold is used ultimately as a raw material. On the other hand, the countries of Western Europe, particularly France, wanted to make it legally possible for gold to take part in settling balances of payments at market prices, which at that time were about four times higher than the old official price of 35 dollars an ounce. As a result, in November 1973 the heads of the central banks of the main capitalist countries agreed that they could, if so desired, sell gold from their reserves on the free market.

Taking the initiative and possibly trying to set an example to other countries, the US government at the beginning of 1975 held two gold auctions, selling about 39 tons of the metal. The buyers were exclusively large dealer firms. No other country followed suit, however. Some countries, in particular Italy, were, it is true, using other methods of unfreezing their gold reserves: international loans against gold as security at its market price. But, unlike American practice, in this case the gold does not go into the private, non-monetary sphere, but remains in circulation among central banks.

The auctioning of gold acquired great economic significance, when the International Monetary Fund began to implement the decisions passed in the summer of 1976 at the Jamaica conference. Over four years it had to sell 25 million ounces (777.5 tons) of gold previously contributed by member countries. It was decided to do this by auctioning, and the Fund held the first auction on 2 June 1976, selling 780,000 ounces (24.3 tons) at an average price of 126 dollars an ounce. The number of bidders who offered a high enough price for their bid to be successful was twenty in all.

The IMF auctions attracted considerable attention as a kind of monetary entertainment. There was no special need for these entertainments. Gold could just as well have been sold in the usual way on the market, as it had been during the period of the Gold Pool, for example, in 1961-1968. The method of auctioning was chosen to demonstrate that gold had become an ordinary commodity. In this way the sales of gold from central reserves received the maximum publicity. The Fund publishes the most precise details on the number of successful bids, the amount of gold sold and the prices. The gold auctions were held within the framework of the policy aimed at driving gold out of the international monetary system.

The Washington auctions of the International Monetary Fund are not at all like the traditional noisy, excited gatherings with buyers' shouting and the auctioneer's hammer. The buyers tender their bids in sealed envelopes, indicating the amount they wish to buy and the price offered. The minimum bid is 1,200 ounces (about 37.3 kilograms). So it is obvious that this business is not for the man in the street. In fact bids are usually for much higher amounts than this minimum.

The IMF auctions were first held every six weeks, then every month. The amount of gold sold at each auction was reduced from 24.3 tons to 16.3 tons, and then to 14.6 and 13.8 tons. This was connected with the fact that the Fund set aside a certain amount of gold for preferential sale to the developing countries. The difference between the high auction price and the nominal price at which the gold was calculated on the IMF accounts (35 SDR units an ounce) was put into a special Trust Fund. This money was distributed according to a certain formula among the developing countries which are members of the IMF (except for the oil-producing and some other "wealthy" countries). Each country could use part of its share to buy gold from the IMF. By this programme the IMF sold 46 tons in all, of which more than half was bought by India.[22]

The price at which the IMF sold gold throughout the four-year programme fluctuated sharply in close correspondence with the current market price. The lowest price (109.4 dollars an ounce) was at the auction in September 1976, and the highest (712.1 dollars) at the auction in February 1980. The very size of these fluctuations is an indication of the instability of the market, the turbulent political and economic events that provided a background for the sale programme.

At each auction the total number of bids made by potential buyers was higher (sometimes four or five times) than the amount of gold being auctioned. Therefore only some of the bids were successful. The number of successful bidders on no occasion exceeded twenty and on several occasions dropped to five. In the list of successful bidders we find such well known gold dealers as the Big Three Swiss banks, members of the London market led by the house of Rothschild, the West German Deutsche Bank and Dresdner Bank, and big American firms, first and foremost, the Engelhard company. Up to 1978 the central banks could not formally take part in the auctions. But France, which invariably has a special position on the gold question, bought a certain amount of the metal through an intermediary, the Bank for International Settlements, an international institution located in Basle.

In 1978 the central banks obtained the formal right to buy gold on any market at any price. This did not affect the auctions, however: as before they were attended almost exclusively by private dealers.

The experts take the view that at the 1979 auctions these firms were buying gold largely on behalf of "oil sheikhs". The London *Economist* wrote: "Swiss (and West German) secrecy has cloaked the true identity of the purchasers at the regular IMF and United States treasury auctions. But when a single German bank snaps up all the gold on offer, it does not take too much figuring to conclude that Middle Eastern clients are probably the only people left in the world with the cash to back such instructions."[23]

The IMF auction programme ended in May 1980. Long before this specialists and the financial press had begun to discuss whether it would be continued. Already in 1979 the opinion was unanimous that there would be no further sales of gold from the remaining reserve of the IMF (about 3,200 tons), at least in the next few years. By the time the sales were completed this had become perfectly clear.

In 1976-1977 it looked as though the Fund would gradually sell off all its gold. But in four or five years the situation has changed. Gold has again showed its effectiveness as a central reserve of international liquidity. It is playing a significant role in the European Monetary System. New ways are being discussed of using the IMF gold reserve as a guarantee fund for the intended issue of additional amounts of SDR in exchange for the dollar reserves of member countries. In order to pass a decision on resuming sales the IMF must have a majority of not less than 85 per cent of the votes of the member countries on its steering bodies. Even if this proposal were put on the agenda, it would not get that many votes today.

The developing countries received certain benefits from the IMF sale of gold. The total of the above-mentioned Trust Fund, from which they obtain sums in various forms, was about 5 billion dollars. Taking this into account, they would be interested in the sales continuing. However, their influence in the IMF is not great, and there is no unity among them.

In the United States the question of the sale of gold from the national reserve is the subject of constant debate and dispute. There are influential circles supporting the sale of gold. These are, first and foremost, consistent supporters of the demonetisation and final removal of gold from the monetary system. The position of the pragmatists, who believe that with the deterioration in the position of the dollar and the drop in its exchange rate the sale of gold for dollars may be an effective way of supporting and stabilising the dollar, is slightly different.

The opponents of "squandering" the gold reserve also enjoy considerable influence. They argue that in any case the US gold reserve is now almost three times lower than its highest level. At a time when other countries are holding on to their gold and using it as a reserve for international payments, the United States should take care of its gold reserve too.

It is well known that the Pentagon has always opposed the sale of gold. This extremely influential department and the political circles connected with it regard the contents of Fort Knox as a military-strategic reserve. The generals are probably not interested in the fine points of demonetisation, but centuries of experience tell them that at a critical moment gold has never yet done anyone any harm or been superfluous.

As a result the actual gold policy of the USA shows inconsistency and lack of a long-term perspective. Under pressure from the depreciation of the dollar on the world foreign exchange markets the US government in April 1978 decided to hold a series of auctions at each of which 300,000 ounces (9.3 tons) of gold were to be sold. Technically they were similar to the IMF auctions. It was officially announced that these operations had two aims: first, to reduce the balance of payments deficit and support the exchange rate of the dollar; and, secondly, to confirm the desire of the United States to continue the process of abolishing the international monetary functions of gold.

As we can see, the demonetisers were pressing their cause. The reaction in Western Europe to the American actions was not very favourable, and frequently sarcastic. The French *Figaro* carried an article by its financial commentator under the headline "The Bluff of the American Gold Sales". The author called it a diversionary operation by the United States, which was selling gold worth several hundred million dollars and at the same time forcing tens of billions of paper dollars (the self-same debt-reserves) on other countries. The USA was putting on one of its "shows" instead of "occupying itself with its fundamental problems".[24]

When these sales had no appreciable effect on the position of the dollar, their scale was considerably increased: at first up to

750,000 ounces (23.3 tons), and then even up to 1.5 million ounces (46.7 tons) at each auction. With such a scale of selling the American auctions eclipsed for a while the more modest regular IMF auctions. Subsequently the scale and tactics of the auctions have changed frequently. Throughout the whole of 1979 the US government sought to bring down the inexorably rising curve of the price of gold.

That influential paper of American business circles, the *Wall Street Journal,* explaining the motives and logic of the authorities' actions, wrote that gold dealers might wonder why the government was so interested in their market, if it really believed that gold was nothing more than merely a commodity. In fact it did not believe this. The connection between gold and the dollar was more or less broken. American currency did not drop automatically when the price of gold rose. But a rising price—especially if it soared sky high—was a very clear indication that investors wanted to move from paper money to some more tangible assets. It was a barometer of inflationary expectations and itself a factor of increasing inflation. A rapidly rising price tended to arouse doubts in people's minds as to the value of paper money, which most people possess, and showed a great lack of confidence, the journal concluded.

The market snapped up American gold without any signs that it had had enough. After selling about 500 tons in 1978 and 1979, the US government stopped the auctions in November 1979. As mentioned above, this sudden change in policy added fuel to the flames of the gold rush that had broken out at the end of the year and caused the price of gold to soar to fantastic heights.

For various reasons some other countries which do not produce gold have been forced in recent years to sell part of their reserves on the free market. This is done, of course, not because of some doctrinaire desire to promote demonetisation. A certain amount of gold was sold by Portugal which is experiencing great economic and financial difficulties. The government of India, in an effort to economically undermine the smuggling of the metal into the country, which recent estimates placed at 60 tons a year, is selling gold from its own reserves on the domestic market.

Economists with opposing views on the economic functions and future of gold can interpret the facts mentioned above in completely different ways. Those who regard the demonetisation of gold as the main trend argue that sales of gold at sharply fluctuating market prices point to the turning of gold into an ordinary commodity and the reduction of its monetary functions. Conversely, those who oppose demonetisation regard the extension of operations with gold and the rise in its price in paper money as proof of the retention and increase of its monetary functions. These divergencies of opinion show at least that the problem of the economic role of gold is a most acute and at the same time a rather complex one. They reflect the fact that gold retains an important position in the capitalist economic system.

IX. THE FATE OF THE YELLOW METAL

Gold is going through a critical stage in its long history. This is connected with the development of the state-monopoly stage and, particularly, with the evolution of the monetary system and international economic relations of present-day capitalism. As we have seen, complex and diversified processes, permitting of different interpretation, are taking place in the economic functioning of gold. Two questions which will largely determine the future of gold are of special interest: first, how will its purchasing power develop in relation to commodities (or, to be more precise, in relation to *other* commodities); and, secondly, what will be the ratio of the monetary and non-monetary functions in its economic life. These questions are closely interconnected. In the preceding chapters the foundation has been laid if not for providing unconditional answers to these questions (that would be too presumptuous), then at least for outlining their possible development.

Value, Price and Purchasing Power

Historically their interrelationship has always been the nucleus, the main content of monetary theory. With the ousting of gold from the monetary system and the divorce of paper-credit money from the gold basis, the centre of gravity of this theory has shifted. The main place in it has been taken by such questions as the influence of money and the monetary system on economic growth, their role in state regulation of the economy and the fight against cyclical crises. But the meaning and interconnection of the value, purchasing power and price of gold continue to be an extremely complex problem.

At one time the phenomena of metal money and its circulation played a decisive role in the formation of the science of political economy itself. Observations in this sphere served in many respects as material for the first attempts at economic analysis. The sharp drop in the value and purchasing power of gold (and silver) in the 16th and 17th centuries—the so-called *price revolution*—provided ample material and served as the basis on which this new science developed. As the eminent Soviet expert on money I. A. Trakhtenberg remarked, "it is no exaggeration to say that economic science grew out of the question of money".[1] A similar opinion had been expressed earlier also.

The phenomena of metal circulation may, however, seem simple compared with what happens in economics with the emergence and evolution of paper-credit money. Where does its purchasing power come from and what determines it? This question is of both theoretical and practical importance.

In Marxist economic science during recent decades there has been active discussion of the relationships between paper-credit money and gold and of whether, as in the 19th century, it should be seen only as representing gold or whether, with the development of 20th-century capitalism, it is becoming divorced from gold and acquiring the ability to measure the exchange value of commodities independently. What comes first in modern inflation—the depreciation of paper-credit money in relation to gold or its depreciation in relation to commodities? This question in various interpretations and aspects is occupying the minds of economists.

The puzzle of problems of money and gold has been the subject of many witty and ironical remarks. The founder of the Rothschild banking house and dynasty is reported to have said once that there were only two men in the whole world who understood gold. One was an obscure clerk in the Bank of France, the other a director of the Bank of England. "Unfortunately," he added, "they disagree."[2]

But let us turn to the heart of the matter. The basis of our economic thinking is the *labour theory of value*. It was created in the 17th and 18th centuries by the classics of bourgeois political economy and developed and transformed by Marx.

Gold shares one very important common feature with all other commodities: it is produced by human labour. Labour creates the value of gold, as of all commodities. The amount of value is determined by the amount of labour required under normal conditions of production to produce a commodity unit, in this case to produce an ounce or a gramme of gold. A rise in labour productivity reduces this amount (in other words, working time measured in man-hours or other similar units) and, consequently, reduces the value.

But value is concealed somewhere in the depths of the economy. It can manifest itself outwardly only in the form of exchange value, i.e., a system of ratios in which a commodity is exchanged for other commodities. This system of ratios is called the purchasing power of gold. Basically the purchasing power of gold (like the exchange value of any commodity) depends on the ratio of labour expenditure on the production of a unit of gold to that on the production of other commodities. Hence it follows that if labour expenditure on the production of gold is cut more quickly than on the production of other commodities(in other words, labour productivity in gold production grows more quickly), all other things being equal, the purchasing power of gold should drop. And, conversely, if labour productivity rises more quickly in other industries, the purchasing power of gold should tend to grow.

The matter is considerably complicated by the fact that the gold-mining industry is one of those industries where a physical factor, the richness of deposits, has a decisive influence. At one and the same time gold is mined at more rich and less rich deposits, and the number of these deposits is limited by the workable geological reserves of the metal.

All deposits being worked can be put on a sliding scale ranging from the richest field with the highest labour productivity and the lowest individual value of the product to the field where, by virtue of poor ores and bad conditions of the deposits, labour productivity is lowest and individual value highest. But society needs the produce of all these fields, including the relatively poor ones. Naturally, certain economic factors determine where the dividing line lies between a marginal (poor but nevertheless workable) field and a sub-marginal field which is not workable. Society does not, so to say, sanction the working of such an unproductive field. But the labour expenditure on the marginal field becomes determining for the whole industry. They constitute the basis for forming the value of gold. This means that the owner of the field, a private entrepreneur, can recoup his costs and obtain a satisfactory profit on his invested capital. The owners of all fields better than the marginal one obtain not only a satisfactory profit, but also a certain surplus which from the point of view of political economy is *rent*. In this way under capitalism private owners of relatively richer fields derive a benefit from their property. If it is sold, the selling price contains a rent previously calculated and added up for a number of years. The buyer pays this price only if he is certain that he will receive the rent regularly. This rent is called *differential* rent in Marxist political economy, because it is conditioned by the natural differences in the wealth and location of fields. But the owner of a marginal field (the worst of those exploited) is also a private owner and wishes on economically legitimate grounds to receive an income from his property. Society is forced to give him this income, on account of which the value of gold is even higher. This additional amount of value is called *absolute rent*.

Rent from gold mines and fields is basically no different from ground (agricultural) rent and is subject to the same general laws. At the same time economically the mining of gold differs considerably not only from agriculture, but also from other branches of the mining industry. Therefore the formation of the value of product shows some specific features. The most serious attempt with which I am familiar to apply the labour theory of value and the theory of rent to the problem of gold is to be found in a book by the Soviet economist S. M. Borisov. However, it too leaves much unexplained. In addition, the author's efforts were directed largely to explaining the process of the formation of the value of gold and related phenomena in the specific conditions of the fixed price of gold which existed until the end of the sixties. Today, as we know, the situation has changed and the price of gold is determined directly by the market.

The law outlined above is merely the most abstract scheme of the formation of the exchange value (purchasing power) of gold. In economic reality it is formed in the most complex way under the influence of many factors operating both directly in the production of the metal and in the economy as a whole.

What are labour expenditure and labour productivity? They

are formed of two components: live labour applied directly in the mining of gold, and embodied labour—equipment, machinery, transport, electricity, etc. Labour productivity is the aggregate productivity of both its forms. It depends on many factors, primarily on the richness of the deposit, the technological progress and the skills of the workers. But it turns out that labour expenditure and productivity also depend on how the gold is sold, on its purchasing power. There is not only a direct link, but also a feed-back between labour value and purchasing power. This feed-back operates through the mechanism of determining the marginal conditions of producing the metal.

It is well known that the purchasing power of gold can fluctuate considerably without any significant changes in the conditions of its production, under the influence of such factors as the phase of the economic cycle, changes in the spheres of use of gold, government policy, etc. Let us assume that under the influence of a set of similar factors the purchasing power of gold has dropped. A real historical situation of this kind was the period of the 1950s and 1960s. In these conditions the producers, which today means large companies, abandon poor deposits, close down mines and fields with low productivity, and concentrate on rich deposits. S.M. Borisov describes the course of events as follows: "Production of the metal is concentrated on the more productive fields, and now it is the poorest of these that will determine the new value of gold. Since the expenditure of labour on these new poor fields will be less than on the old poor ones, the value of gold will drop."[3] Obviously this may cause a decrease in rent from the relatively better fields, since the difference between the best and the worst is reduced. This was in fact observed in South Africa at the time and aroused great dissatisfaction among gold producers with the low price of gold which was determined at that time largely by non-market factors.

The reverse phenomenon was observed in the seventies. A considerable rise in the price and purchasing power of gold enabled companies to shift to the mining of poorer rock, even including the reworking of dumps formed during previous mining. Evidently value in the industry began to be determined by the production conditions in poorer areas than before and, as a result of this, it rose. This, as it were, gradually provided a sort of basis for the high market price and purchasing power of gold.

But what would have happened if new rich deposits of gold were found? Their involvement in production would turn the relatively richer areas into marginal ones, the value of gold would fall as a result of the reduction in labour expenditure and, all other things being equal, this could bring about a drop in its purchasing power. It is important to stress this to make it clear that production conditions and not the market still lie at the basis of the whole process.

The purchasing power of gold depends directly on the price of gold in paper-credit money and on the price level of all other commodities. It may be compared with the concept of a real wage which depends, on the one hand, on the nominal (money) wage and, on the other, on the prices of the commodities that are bought with this wage. The price of gold (like the wage) may be expressed in monetary units, for example, in dollars. But it should be compared with the prices of a countless number of commodities. This makes sense only dynamically and is done with the help of indexes. For example, the price of gold at a certain period was 200 per cent in relation to a certain basic period, and the price index of other commodities (i.e., the average rise in all prices) was 150 per cent. Dividing the former by the latter we see that the purchasing power of gold was 133 per cent of the basic level, or rose by 33 per cent. In other words, in that year it was possible to buy 33 per cent more commodities with the same amount of gold than in the basic year.

In times of the gold standard the price of gold was stable. It was not a price, in fact. For a price is an expression of value in money. But if gold is money, it means that gold expresses its own value in itself. The price of gold was merely the inverted gold content of the monetary unit.

The prices of all commodities were, of course, expressed not in ounces or grammes of gold, but in dollars and other national currencies. These prices were mobile and changed constantly. The price of gold, conversely, was fixed. Therefore the purchasing power of gold could change only through a change in the general level of commodity prices. A rise in this level meant a drop in the purchasing power of gold, and vice versa. In the 19th century the general level of prices and its movement were already measured with the help of price indexes, although at first they were very unreliable.

Under the gold standard selling gold meant simply exchanging bars for coin or for banknotes exchangeable for coin. The buying and selling of gold meant basically only a change in its form. But when paper-credit money is not exchangeable for gold and is the only real form of money in circulation, the price of gold in this money no longer differs in principle from the prices of other commodities. The buying and selling of gold becomes very similar to the buying and selling of any commodity. The producer of gold, in selling it for dollars or pounds sterling, is no different in principle from the producer of diamonds, copper or lead. Today this is particularly obvious.

The history of the last few centuries contains two periods when the purchasing power of gold changed sharply and spontaneously under the influence of changes in the value of gold.

The first was the period of the price revolution mentioned above. A stream of "cheap" gold and silver from America flooded into Western Europe. Lower labour expenditures began to determine the value of the precious metals. In Spain, where the effect of the stream of precious metals was most direct, the average level of prices rose 3.5 times between 1500 and 1600.[4] Somewhat later the same thing happened in other European countries. This meant a considerable, although by modern standards rather slow, fall in the purchasing power of gold and silver. The most reliable statistics beginning from 1560 are for England. According to them, over the 1630s the average purchasing power of gold was two-thirds of its purchasing power in the 1560s.[5] Probably a certain drop in the purchasing power of gold took place before 1560, but because of the chaos in the coin system at that time serious estimates are impossible.

The second period of interest to us is the last three decades of the 19th century, the period of the establishment of the gold standard. From 1872 to 1896 the wholesale price index in the USA dropped by 50 per cent, in England by 39 per cent, and in France by 43 per cent.[6] This would seem to be explained as follows: during this period no new rich deposits were discovered (gold mining in the south of Africa was still on a small scale), and labour productivity in the gold industry was not rising, possibly even falling. Whereas in most other industries there was rapid technological progress and labour productivity was rising. Hence the rise in the purchasing power of gold.

The growing cost of gold had as one of its consequences a rise in the profitability of gold mining. By accepting an unlimited amount of newly-mined gold at a fixed price, governments were, so to say, subsidising the gold-mining industry. Probably this unlimited demand in itself produced a rise in the value of gold by the following mechanism. Poor ores and placers where labour expenditures per unit of the product were high were also drawn into production. Insofar as the areas for mining gold in the world are limited and production cannot be freely increased, these poor conditions of mining determine its value.

The purchasing power of gold was unusually high by the end of the century and possibly reached its highest level for three centuries. In the following period it exceeded this level only in the 1930s, when the gold content of the pound and dollar was reduced spasmodically and the price of gold rose accordingly, and also in our days, the late seventies and early eighties.

By the end of the 19th century an ounce of gold could purchase, roughly speaking, twice as many goods as thirty years earlier. Profits in the gold industry were extremely high and the hunt for gold was particularly furious. This may explain partly the terrible intensity of the gold rushes of that period.

After 1914 the position with gold changed. Its price in the paper-credit money of the different countries ceased to be fixed. The purchasing power of gold now depended on two factors: commodity prices in paper-credit money and gold price in the same money. This price became an instrument of government policy, an instrument of state-monopoly regulation. The actions of the US government were of special economic and political significance.

In March 1933 Roosevelt's government which had just come to power abolished the gold standard and introduced, to use the modern terms, a floating gold price and floating dollar rate. The state, as the monopoly buyer of gold, gradually raised the price. It is said that the President sometimes fixed the price of gold for the day at breakfast in a chat with the Secretary of the Treasury, Henry Morgenthau Jr., and the head of the Reconstruction Finance Corporation, Jesse Jones. One day the price was put up by 21 cents an ounce only because Roosevelt thought seven was a lucky number, and multiplied it by three to get a reasonable amount.[7] By January 1934 the price had reached 35 dollars an ounce and was fixed at this level. It was almost 70 per cent higher than the old price.

The whole operation was carried out on the assumption that the price of gold would pull commodity prices up, stimulate demand and help to drag the economy out of the morass of a severe crisis. Commodity prices hardly reacted at all to the rise in the price of gold, however. Insofar as the price of gold rose sharply and commodity prices did not change, there was a rise in the purchasing power of gold. Unlike the events at the end of the 19th century, this was the result of an intentional act, a political decision which had nothing to do with a change in the value of gold.

One of the consequences of the rise in the purchasing power of gold was a considerable growth of its production. This is explained by the fact that in the concrete conditions of the 1930s the South African companies not only went over to working poorer ores, as in the seventies, but rapidly built and put into operation new mines. The mining of gold in the thirties was less concentrated in South Africa than today, and in the specific conditions of other countries production frequently reacts more strongly to price changes.

From 1934 to 1971 the official price of gold was supported by the US government at the constant level of 35 dollars an ounce. Although at times the free market price was higher, the latter was of limited importance then *(Table 6)*. During the period of the Gold Pool it did not differ from the official price at all. Meanwhile, however, under the influence of wartime and post-war inflation the wholesale price index almost tripled in the USA during this period. There was a sharp drop in the purchasing power of gold, which by 1970 had reached its lowest level in the history of the United States, not counting a short period in 1918-1920.[8]

The depreciation of gold did not have enough real economic foundations and was to a very large extent the result of the influence of state policy on the economic sphere.

In persistently keeping the price of gold unnaturally low, the United States was guided first and foremost by political motives.

Table 6

Purchasing Power of Gold in the USA
(annual average)

Year	Price of gold in dollars per ounce*	Gold price index	Wholesale price index	Purchasing power index**
1913	20.67	100.0	100.0	100.0
1929	20.67	100.0	136.4	73.3
1934	34.81	168.4	107.3	156.9
1940	35.00	169.3	112.6	150.4
1950	35.00	169.3	227.2	74.5
1960	35.00	169.3	263.6	64.2
1970	35.00	169.3	306.6	55.2
1971	40.81	197.4	316.3	62.4
1972	59.14	286.1	330.7	86.5
1973	97.22	483.8	374.1	129.3
1974	162.02	781.7	444.6	175.8
1975	160.96	775.7	496.4	156.3
1976	124.82	601.5	542.8	110.8
1977	147.72	711.9	567.4	125.5
1978	193.24	931.3	602.1	154.7
1979	306.67	1,483.6	654.3	226.7
1980	612.59	2,963.7	745.9	397.3
1981***	482.88	2,336.2	805.6	290.0

* Up to 1970 the official price, after 1970 the free price of the London market.
** Obtained by dividing the price of gold index by the wholesale price index.
*** First half.

Sources: *World Currencies*, Finansy, Moscow, 1976, 1981 (in Russian); *International Financial Statistics*, November 1981.

It was thought that a rise in the official price of gold (a devaluation of the dollar) would undermine the prestige of the United States.

In 1971-1973 the United States was forced to abandon its position. This took place not on the basis of impartial statistical estimates and calm discussion, but in an atmosphere of acute crises, panic and emergency. The official price of 42.22 dollars an ounce established in 1973 had no real economic significance right from the start. The market price rose four times between the end of 1971 and the end of 1974. According to the wholesale price index in the USA the purchasing power of gold in 1974 was 76 per cent higher on average than in 1913, and 12 per cent higher than in 1934. In paper dollars, the price of gold was almost eight times the 1913 price.

As often happens with commodities of the speculative type, this rise was followed by a drop in the price of gold, which in 1975-1976 took place against the background of a rise in the general price index. The purchasing power of gold again dropped considerably and did not rise above the 1974 level until 1979. In 1980 the purchasing power of gold reached the highest level in history, but later fell considerably.

Throughout the last fifty years the movement of the purchasing power of gold has taken place against a constant rise of commodity prices. Individual commodities may sometimes get cheaper, but the general level of prices is rising practically all the time, which now appears as a kind of law of nature. The purchasing power of gold is formed as a result of the constant race of the price of gold and the general level of prices.

Past and Future

When Roosevelt raised the price of gold in 1933, this was considered an important act of economic policy. Even in 1960, when the "gold rush" happened and the free price of gold soared to 41-42 dollars an ounce with an official price of 35 dollars an ounce, the United States sounded the alarm and soon organised the Gold Pool to regulate the market and the market price. Now things have changed. In the summer of 1973, when the price passed the "psychological barrier" of 100 dollars (which now seems ridiculously low), Mr George Shultz, the Secretary of the American Treasury, said that he was more concerned about the price of hamburgers than of gold. This joke concealed a fair amount of hypocrisy and posing. That old sceptic, the London *Economist,* which has existed and enjoyed authority since the time of Marx, wrote, "But not many people believed him." And explained that the further the market price of gold has moved away from the official price the greater the loss of credibility for the dollar.[9] This applies to the present too. Nevertheless there is a grain of truth in the Secretary's joke: the price of gold has not lost its importance as an indicator of the international standing of the dollar, but people feel inflation not by the price of gold, but by the price of meat and other essentials, and this is of major political importance.

Insofar as gold had ceased to be the basis of monetary systems its price in paper-credit money lost its former significance. Actually it would be more correct to say that it has not lost, but *changed* its significance. It is not so much a direct element of the monetary system, as an indicator of the general economic, financial and political situation in the capitalist world.

The world price of gold is expressed in dollars. This reflects the role of the US currency in the world capitalist economy. Let us recall that on the London market, which retains the importance of the main world market, gold is quoted in dollars. The price of gold may, of course, be expressed and actually exists on the market in many other currencies, but these prices express only the relation of a given currency to the dollar. A currency's exchange rate depends in the final analysis largely on the rate of inflation in the country in question. If this rate is considerably less than in the USA, the exchange rate of the currency for the dollar may rise. In this case the price of gold in the given currency may rise more slowly than in dollars, and may even fall. This was the situation in 1976-1978, when the exchange rate of the dollar dropped considerably in relation to relatively stable currencies such as the West German mark and the Swiss franc. On average in 1978 the dollar price of gold was 20 per cent higher than the average for 1975. However, the price in marks was 3 per cent lower than this level, and in Swiss francs as much as 17 per cent lower. In 1979, on the contrary, the rise in the price of gold was almost identical: compared with the previous year it rose on average by 58 per cent in dollars, 46 per cent in marks, and 48 per cent in francs. The differences in the behaviour of prices at different periods reflect an important fact. In the first period the main cause of the rise in prices was, so to say, the individual weakness of the dollar, in the second the loss of confidence for paper money in general, the flight of money capital into gold as a stable value.

The rise in the dollar price of gold is a direct reflection of the inflationary depreciation of the national currency of the USA. But insofar as the dollar is used extremely widely in world trade, gold is becoming an indicator of the global depreciation of the dollar. This may be illustrated with the example of oil: over the last ten years the dollar price of oil rose 12-15 times and even more on the world market. This is, of course, connected with the fall in the purchasing power of the dollar within the USA, but it is also caused by specific factors not directly linked with domestic inflation in the USA. The same may be said of gold.

It is interesting that the dimensions and even the timing of the sharp rises in the prices of oil and gold were close to each other until 1981. On a graph the two curves run side by side. An empirical "pattern" has even been established, according to which one ounce of gold "should" cost 17 barrels of oil (a barrel is equal to 159 litres—the usual measurement for oil). In 1971, before the "price revolution" of the seventies, an ounce of gold cost 35 dollars on the world market, and a barrel of oil about 2 dollars. In 1980 the respective figures were 500-600 dollars an ounce and 30-35 dollars a barrel.

Such a coincidence cannot be regarded as the manifestation of a long-standing and stable pattern. The factors that produced the price rise in both commodities have a great deal in common, but they also differ essentially.

Insofar as we shall be discussing forecasts on the following pages, we shall make use of the material of eminent futurologists and forecasters. One of the most authoritative long-term forecasts has been prepared by a group of United Nations experts led by Wassily Leontief. The authors of the forecast were not interested in the prospective price of gold, since its economic role as an industrial raw material is insignificant. But in relation to the price of oil they concluded that its relative price (i.e., the quotient from dividing the oil price index by the general price index—in fact, the purchasing power of oil) will rise 2.2 times by 1990 compared with 1980, and 2.5 times by 2000.[10]

Inflation is built into the economy of modern capitalism. Any forecast of the development of the capitalist economy assumes that the growth in the average level of prices will continue in the foreseeable future. The question is whether or not the rise in the price of gold will exceed the average level of dollar commodity prices. In other words, whether the purchasing power of gold will rise or fall.

The final basis of the price and purchasing power of gold is its value, the movement of the labour expenditure on mining compared with the labour expenditure for the production of other commodities. But this law works its way through the market, through the relation of supply and demand. Therefore in order to forecast the purchasing power of gold one must study the specific conditions of its production and consumption.

Among Western books on gold published in recent years a work by the American scholar Roy W. Jastram entitled *The Golden Constant. The English and American Experience, 1560-1976* deserves attention. It is an academic study of the movement of the purchasing power of gold in relation to commodities over more than four centuries. History, economics and statistics form an interesting blend in this book, which is essential to any study of the history of gold. Four centuries of experience cannot fail to be of interest for forecasting also. The conclusions of the author may be summarised as follows.

1. Fluctuations in the purchasing power of gold over the last four centuries were relatively small, and it stubbornly kept returning after deviation to a fairly stable constant figure. The above-mentioned significant changes (1870s to 1890s and 1930s) were compensated relatively quickly by movements in the opposite direction. The author quotes the following interesting series showing, on the base of 1930 taken as 100, an index of the purchasing power of gold in Britain over fifty-year intervals, beginning with 1600 selected arbitrarily: 1600—125; 1650—97; 1700—120; 1750—111; 1800—76; 1850—111; 1900—143; 1950—103.[11] As we see, the deviations from 100 are not too great, and eventually after 350 years the purchasing power of gold returns to the 1600 level.

2. Of the two factors that determine the purchasing power of gold, the commodity price index was far more mobile than the price of gold. It was the movement of this index that invariably returned the purchasing power of gold to a constant level. In Britain the price of gold remained practically unchanged from the end of the 17th century to 1931 (with the exception of short periods during the Napoleonic Wars and World War I). Over these approximately 250 years the commodity price index dropped to its lowest level of 63 (in 1896) and rose to 259 (in 1920).[12] In the USA the price of gold remained almost unchanged from 1800 to 1933. The lowest point of the commodity price index was 54 (1896) and the highest 179 (1920).[13] It is interesting that the years of the extremes in both countries coincide. After rising with a leap in 1930s the price

of gold in Britain and the USA remained unchanged for several decades, whereas the level of commodity prices changed sharply almost entirely on a rising trend.

3. In periods of inflation (this is what the author calls all periods of price increases) the rise in the price of gold lagged behind the rise in the commodity price index, as a result of which gold depreciated. Not without reason the author supposes that this will be the most unexpected conclusion of his study. And, indeed, it is usually thought that investing money in gold is the best way of insuring against inflation. It transpires that up till the 1970s this was not the case. During the Napoleonic Wars the price of gold in Britain rose by a maximum of 41 per cent, and the commodity price index by 59 per cent; during and after World War I the respective figures were 33 per cent and 195 per cent.[14] A similar situation can be seen in the USA during the Civil War of 1861-1865 and World War I. The loss in the purchasing power of gold during the period of inflation of the 1940s to 1960s has been discussed above.

4. The purchasing power of gold rose in periods of deflation (a fall in commodity prices). The price of gold did not change in such periods and at the beginning of the 1930s even rose with a leap while commodity prices dropped. This conclusion is logically in line with the previous one.

If we were to try and transpose this experience automatically on the future, we would obtain the following forecast: the purchasing power of gold, which reached an all-time peak in 1979-1980, will diminish in the next few decades in order to return to its traditional level; this will happen not so much through a change in the price of gold as through a change in commodity prices; since there can be no doubt that the coming years will be a period of inflation the rise in the price of gold will lag behind the rise in commodity prices.

True, in 1981-1982 the purchasing power of gold diminished abruptly owing to a fall in the price of gold, while the general level of commodity prices continued to rise. This, however, reflects mainly the operation of short-term factors and cannot determine the trends of development for a longer term.

But in this case historical experience cannot be regarded as decisive for the future. Jastram himself indirectly agrees with this when he writes: "Thus it is easy to predict a more nearly free market for gold in the future, with the attendant possibility that gold will become a better hedge against inflation than it has proven over past centuries."[15]

The period of the 1970s and 1980s differs fundamentally from all inflationary periods in the past. In all the cases mentioned we were dealing with comparatively short periods of wars and upheavals which were followed by the stabilisation of monetary systems and the restoration of the gold standard. Today we have inflation of an "organic" kind, which is developing in peacetime and does not promise any stabilisation. We do not have the historical experience to predict the behaviour of the purchasing power of gold in these conditions. In most cases in former periods of inflation the price of gold had been artificially prevented by the state from rising as dictated by the market. This was a strong factor in the subsequent deflation. The classic example is the experience of Great Britain in the 1920s and the return to the gold standard by Churchill. In the 1940s to 1960s the world dollar price of gold was also supported at an artificially low level, but in the new conditions which developed after World War II this could no longer end with deflation and a drop in the general level of commodity prices, but led to the renunciation of a fixed price for gold and a multiple rise of the market price. There can hardly be a question of a fixed price for gold in the future.

Many of the patterns of the formation and movement of the prices of minerals that have emerged in the last decade (this question has already been touched upon above in connection with the price of oil) apply to gold. UN experts believe that in the 1980s and 1990s the most important factor in the rise of relative prices not only of oil and natural gas, but also of nonferrous metals (copper, nickel, zinc and lead) will be the drawing into operation of less productive and more expensive sources.[16] It is probable that this will also happen to gold.

The total amount of gold in the earth's crust and the ocean water is immense. It is estimated in millions of tons. But industrial (economically accessible at the present level of technology) reserves are far smaller. According to the estimates of the US Bureau of Mines, the industrial reserves of gold in the capitalist world are about 35,000 tons, including 25,000 in South Africa. At the present rate of output it will last for thirty or forty years. These estimates were based on 1974 prices which were about 200 dollars an ounce. With higher prices the estimate of supplies may increase, because the working of poorer deposits becomes profitable.

However, the natural reserves of gold tell us little about the prospects for mining it. Of decisive importance is the discovery of new rich deposits capable of providing a considerable increase in output. Does such a prospect exist? At least not in the near future. The most recent big discoveries in South Africa were made in the 1950s and have long since been creamed off. Geologists do not promise sensations. And even if some relatively rich deposits were found, it would take a long time to work them. It would be ten years before such finds could influence the gold market. According to 1976 statistics, only three comparatively small mines were being constructed in South Africa. The companies are not investing their growing profits into the construction of new mines. Apart from geological and economic factors this is explained by the political situation: the crisis of the South African racist regime and the upsurge of the African national liberation movement. An indirect indication of the not so good prospects for the gold-mining industry of South Africa is the relatively low level of the prices of company shares, which does not correspond to the rich profits and dividends. In normal conditions an investor buys a share at a price which is ten or twenty times the estimated annual dividend. Recently, however, the rates of shares in the gold-producing complexes have barely been five times the annual dividend. This means that gold shares have become a risky investment. The investor is trying to cover his costs in a short period—five years. As *The Economist* remarks, this could mean that the investor of capital does not expect the regime to last for long.[17]

Apart from this, the monopoly position of the companies is of great importance for the gold-mining prospects in the country. Like the oil-producing countries they are not trying to produce as much gold as possible. It is far more profitable for them to keep poor deposits running at full capacity, leaving the richer ones for a rainy day.

The leaders of the South African gold-mining industry are acting on the assumption that up to the mid-1980s the level of output (700-750 tons) will remain more or less constant and after that production will drop. Officials of the Anglo-American Corporation estimate that by the end of the century output will probably be half what it is at present, irrespective of the level of the price of gold. Naturally, this prediction assumes that nothing will change politically in South Africa. And that, of course, depends little on the companies.

Since the Republic of South Africa produces three-fourths of the gold in the capitalist world, this country's prospects largely determine the general picture. It must be added that in other countries a large part of the gold is produced as a by-product, and the size of output depends not on the price of gold, but on the production level of the non-ferrous metals which are the main product. Gold production in this industry will probably grow, but very slowly. In recent years there have been reports from various countries about attempts to reopen abandoned mines and the appearance of new prospectors. Production of placer gold rose considerably in Brazil. In Australia, California and even parts of Western Europe enterprising or naive people are washing gold in the beds of rivers and streams once famous for their gold-bearing sands. But this is a mere drop in the ocean.

The possibility of the discovery of new rich gold deposits cannot be excluded, of course. The absence of such discoveries for at least a quarter of a century does not determine the future fully, particularly because the sharp rise in the price of gold is stimulating prospecting. However, rich gold deposits are extremely rare. At least today there is no authentic information about prospecting which promises to be extraordinarily successful.

As far as one can judge, gold will remain a scarce commodity on the world market in the next decades. Perhaps the most obvious indicator of this scarcity is the huge disproportion between the increase in the overall volume of production in the capitalist world, in the output of other metals in particular, and the output of gold, which is only slightly higher than the 1913 level, far below the 1940 level, and continuing to drop.

As already mentioned, in 1977-1979 the relationship between the supply of gold and the demand for it was determined to a large extent by considerable sales from the stocks of the United States and the International Monetary Fund. In the foreseeable future there is no prospect of gold entering the market from these sources. It is typical that in January 1980 when, at a time of feverish demand and price rises, the heads of central banks of Western countries met to discuss the question of concerted gold sales to make the market "healthier", the Federal Republic of Germany and France were firmly opposed to this. These two countries are respectively the second and third (after the USA) holders of gold in the capitalist world. The inclusion of gold in the European Monetary System makes it even less likely that the rich countries of Western Europe, which possess large stocks, will sell gold on the market. The reverse is more likely: these countries may undertake in certain conditions to buy gold, something which, as far as is known, they have not done to any significant extent for the last fifteen years.

The private demand for gold is divided into industrial and hoarding-investment, with no clear dividing line between the two. The demand of the jewelry industry reacted comparatively weakly to the rise in the price of gold in 1978-1979. This is undoubtedly explained by the fact that a more important role is now played in its structure by hoarding and investment motives. To put it simply, articles of jewelry are being bought not so much as ornaments, but as means of preserving and increasing value. If the drop in the price elasticity of demand proves to be a prolonged trend, this will have an important effect on the market. In this situation the traditionally high income elasticity of demand, i.e., the feature of jewelry demand to grow more quickly than income, will acquire even more importance. If a family with an annual income of, say, 50,000 dollars gets a rise in income of 10,000 dollars, it will spend, for example, 5 per cent of the 50,000 on jewelry, but 7 or 8 per cent of the extra 10,000. The proportion spent on food will drop. Therefore a rise in incomes of the wealthy strata leads to a more than proportionate increase in demand for jewelry. Naturally, this demand as a whole is a far more complex phenomenon from the economic and social angles.

The jewelry demand still fluctuates depending on the level of prices of gold as a raw material and on the state of the economy. The high price of gold in 1979-1980 and the economic recession caused a drastic fall in that demand. According to some estimates, in 1980 the jewelry industry in the developing countries did not buy any gold at all on the market. As the mid-

70s show, however, a reduction in the price of gold very soon brings about a new wave of demand for jewelry and an increase in the purchases of raw material by the industry. This phase seems to have begun in 1981-1982.

In the future, too, similar fluctuations are likely, but in the long run the jewelry industry is a major and reliable client of both producers and sellers of gold.

The electronics and other industries present a stable, although rather slowly growing demand. This also applies to dentistry and dentures.

The demand of hoarders and investors remains the least predictable. In 1978-1980 it shot up under the influence of a new rise in world inflation and international tension. Obviously a relative drop in inflation and international tension could produce a relative drop in this demand in the future. Statistics, if they are to be trusted in this matter, tell us that in the past there were even years when a dishoarding of gold took place and it left private holdings. However, "inflationary psychology" is now too strong, lack of confidence in paper-credit money too great, and the practice of investing part of one's assets in gold too deep-rooted. More and more social strata and types of gold hoarders and investors who buy gold are appearing. In order to stop this trend some change is needed, which is difficult to imagine given the present socio-economic and political realities of capitalism. Today this trend is taking root, rather than weakening. *The Economist* wrote about the events of 1979-1980: "Behind the speculative froth was a fundamental change in the way people think about gold. No longer was it regarded as an exotic investment but as something that might remain a store of value when paper currencies were suspect."[18] It is easy to imagine what this fundamental change may augur for the gold market.

Every buyer of gold considers, as far as it is possible and useful for him, the main commercial factors important for the transaction. Apart from the price of the metal, the most important of these factors is the level of *interest rates*. For gold not only does not yield the holder an income in the form of interest, but usually requires additional expenditure on storing, insurance, etc. In buying gold the purchaser expects that the profit from the rise in its price will cover the loss of interest which he could have obtained by investing his money in securities or depositing it in a bank account. The higher the interest rates the less profitable the buying of gold. Today this dependence has become even closer than before, because on the gold market a relatively small role is played by the inexperienced hoarder and a major one by the investor and the speculator, who often buy gold with money obtained on credit and are directly affected by the level of the interest rate.

Together with the price of gold at the beginning of 1980 interest rates of the international and national money markets soared to historical heights. Dollar credit cost from 15 to 20 per cent per annum depending on the form, period and collateral of the credit. Only the frenzied pressure of speculation could drive the price of gold up in such conditions: for with such an interest rate loaned capital doubles in four to five years. The fall in the price of gold in January and February 1980 by 300 and more dollars an ounce confirmed this most clearly. Such high interest rates limit the rise in the price of gold.

The relatively low level of gold prices in 1981-1982 is largely explained by the fact that, notwithstanding the economic recession, exceptionally high rates of interest remained on the money markets.

There is a most important factor which can in certain conditions cause a big demand for gold irrespective of the level of interest rates. This is the degree of stability of the international money market itself, which is a system of international banks closely connected with industrial complexes and borrowers from all over the world in a web of credit relations. This system has long had its weak spots, of which the main one is the excessive indebtedness of the developing countries to creditor banks. Suspension of payments by one or, even more so, several leading countries could put the banks in a difficult position. If confidence in a number of large banks were undermined, the whole credit pyramid would shake. It is hard to imagine in detail what would happen in this case, but undoubtedly there would be a frantic rush on gold.

All these points, to my mind, confirm the idea already expressed that the modern position of gold has no analogy in the past.

Gold has always been a mysterious, paradoxical object. As we can see, the situation of recent years contains undoubted paradoxes. The demand for and price of gold were rising, but, contrary to the usual economic laws and common sense, its production was reduced. Gold was demonetised, which has always been regarded as the main threat to its status, but its purchasing power was rising.

The main paradox of the present situation is the following: modern capitalism cannot base its monetary system on gold, but at the same time it cannot manage without it. This contradiction is being resolved in the struggle of different trends of development and is a constant source of tension and bitter collisions. It is of fundamental importance for the evolution of the capitalist monetary system.

Demonetisation or Remonetisation?

The demonetisation of gold is the term applied to the abolition or reduction of its monetary functions. Remonetisation—the reverse process—is the restoration and enhancement of the monetary functions of gold. The question naturally arises as to which of these trends prevails in modern conditions. This question is being constantly argued, because there are facts which testify to the existence of both tendencies. Therefore it may be more correct to ask in what concrete forms and with what consequences these two processes are taking place and what is their relationship.

A Soviet textbook contains the following passage: "Under modern capitalism there has been a growing tendency towards the demonetisation of gold, i.e., the gradual loss by gold of the functions of a monetary commodity... Gold money does not meet the economic requirements of state-monopoly capitalism. It has turned into a kind of 'gold chain' hampering the expansion of the monopolies and the use of inflation as a means of regulating the economy... The process of the demonetisation of gold reflects the level reached in the development and socialisation of production, and also the adapting of the monetary sphere to the needs of state-monopoly capitalism."[19]

In her article the editor of this textbook, L. N. Krasavina, writes: "The struggle around gold has not finished. Although gold has lost its former role under the influence of the demonetisation which is objectively taking place and being accelerated under the pressure of the USA, in recent times there has been a tendency to enhance its monetary role."[20]

The question, to our mind, is put quite correctly here. In the discussions on gold in the West it often looks as if gold could be demonetised or could be allowed to retain the functions of money. Some want the former, others the latter, but the position of the supporters of demonetisation is stronger, which explains why it is taking place. In fact the ousting of gold from the monetary system is an objective process, which does not depend on anyone's will. It has been imposed on capitalism by the very course of historical development. This does not, of course, exclude the fact that the rates and forms of the ousting may be determined by the alignment of forces of influential groups inside the country or in the international arena. It is a well-known fact that the economic and political power of the USA has played an important role in recent years in the formal international demonetisation of gold.

The gold standard corresponded to the conditions of the simpler economy of the 19th century. The inflow or outflow of gold could have an effect on the credit and monetary sphere, the structure of which was also relatively simple, and through it on the economy as a whole. The modern credit and monetary system is notable for its complexity and many elements, and its connections and channels of influence on the sphere of production are equally complex.

The gold standard corresponded to the conditions of pre-monopoly capitalism, when production was dispersed among a mass of small enterprises and none of them individually could influence the prices of their commodities. But it restricts the expansion of large enterprises and their striving to establish high monopoly prices. If prices rose in a certain country, this led to a deterioration in the balance of payments and a loss of gold. Under the gold standard this situation could be corrected in one way only—by restricting bank credit and capital investment and lowering prices. Today this is not acceptable to big capital.

The gold standard is even less compatible with state-monopoly capitalism. It is hard to imagine how capitalist countries could bear vast military expenditure and increase inflation if governments were restrained by the "gold discipline". The different forms of state regulation of the economy that plays such an important role in modern capitalism would also be greatly hampered. The gold standard would limit the strategy of economic growth and anti-crisis measures, because all this is connected with big budget expenditure and credit expansion. Today no government can afford to sacrifice production and employment for the sake of retaining the gold reserve and supporting the exchangeability of paper money for gold.

This is clear even to people who want gold to retain certain limited monetary functions. One most competent such person is the former French president Valéry Giscard d'Estaing. However, he told the National Assembly in 1963: "There are people who believe, and believe in good faith ... that the return to a monetary system based on gold would settle at one blow all the economic and social problems of today. In the final analysis no one can subscribe to this assertion, because this system has existed and has collapsed; there is no regime more inflexible politically, economically and socially than a return to the gold standard. When people speak of adjusting an economic situation, when a disaster hits an agricultural region, what is the word that springs to everyone's lips? It is the word credit. When an enterprise is in difficulty and risks closing its doors, what is the remedy that comes to mind? It is the remedy of credit. How can anyone believe that the system of the gold standard, that is, the suppression of credit, would not show itself as the most inflexible, the most brutal, the most inhuman instrument in relation to the adjustments that have to be made?"[21]

Even the curtailed "gold standard of the Bretton Woods type", restricted only to the international sphere, proved impossible, although deliberate attempts were made to make it more flexible and manoeuvrable. It could exist only in the specific conditions of the two post-war decades, when the USA undoubtedly held the central place in the economy of world capitalism, and the dollar was a satisfactory "substitute" for gold. But when the exchangeability (for foreign central banks) of the dollar for gold and the fixed parity of the dollar began to seriously embarrass the economic policy of the US government, it abolished these vestiges of the gold standard. The formal demonetisation of gold simply recorded the fact that in the course of the monetary crisis the Bretton Woods system collapsed, at least in respect of gold, reserve currencies and fixed parities.

In the present situation there are indisputable signs of an enhancement of the role of gold in the international sphere. But, to my mind, this role is qualitatively quite different from what it was under the classic gold standard and even within the framework of the Bretton Woods system. Gold cannot be the basis of the monetary system if its relationship with paper currencies is not fixed for more or less lengthy periods, if it is not a measure of the value of currencies. And this is the position today. The depreciation and unreliability of paper-credit money will continue to "help" gold and push it to the forefront of the monetary drama. But there can be no return to its former role as the pivot of the whole international monetary system.

Even with the functioning of gold in the international monetary system the question of whether it is money or a commodity in this case is debatable. When Italy received a loan in 1974 from West Germany on security of a part of its gold reserve, the supporters of demonetisation were able to argue that gold had not performed monetary functions here, had not been money. The real money was the West German marks received as credit against the afore-mentioned security. The logic here is obvious: if you take a gold ring to the pawnbrokers and receive a loan for it, you do not regard the ring as money. Anthony M. Solomon, who retired from the post of Under Secretary of the US Treasury for Monetary Affairs in 1980, remarked that governments and financiers were unanimously of the opinion that the remonetisation of gold was impossible. When the question of the possible role of gold in the functioning of substitution accounts within the International Monetary Fund was raised, he said that this would not provide gold with a monetary role.[22]

It is obvious that the monetary elements in the economic life of gold will manifest themselves from time to time fragmentarily, partially and incidentally, but not in the system, not by the restoration of its role as the pivot of the international monetary system. Formal international demonetisation is but a pale shadow of the profound spontaneous changes that have taken place in the economic life and functioning of gold. The essence of money is expressed in its functions. Therefore we must examine to what extent the performance by gold of the functions of money has ceased and to what extent it has changed.

The matter is most obvious concerning the function of a medium of circulation. Gold has not performed this for a long time. Of course, even under the gold standard, it by no means circulated everywhere or in large quantities. But potentially it could always enter circulation. It backed paper-credit money which was only its representative.

Gold has also lost the function of a medium of payment (a medium of settling deferred payments). No one today regards payment in paper-credit money as temporary, not final, or expects to get gold as final payment.

The function of a store of value has changed beyond recognition. The central gold reserves no longer serve as funds for the exchange of paper-credit money. They have lost all connection with internal monetary circulation. In former times private gold stocks played the part of a reserve circulation fund: gold went into them when it became superfluous for circulation, and came out when it was economically necessary to supplement circulation. Today hoards are of an entirely different nature. They serve as an anti-inflationary and political insurance fund and are divorced from money circulation.

However, the initial function of money is the function of a measure of value. The question arises as to how commodity values can be measured and compared if we consider that gold no longer performs this function. When gold was money in the full sense of the word, everything was clear. It measured the value of each commodity and provided the foundation for the price formation for all commodities. If paper-credit money was exchangeable for the metal, it measured the value of commodities in exactly the same way. If there was no exchangeability, paper-credit money could depreciate in relation to the metal. Commodities then acquired double prices—gold (or silver) and paper. But paper prices were not independent, they merely expressed the degree to which this money had depreciated.

In Russia in the first half of the 19th century the main monetary metal was silver and the price of each commodity was expressed in silver roubles. But apart from this there were government banknotes in circulation, a special kind of paper money which was first issued under Catherine II. The state did not exchange banknotes for silver at their face value, and they had the current market rate. Depending on the size of the issues, the political and economic situation in the country and other factors, the silver rouble could cost two, three or four roubles in banknotes (their exchange rate dropped no further than that). Therefore each commodity had two prices—in silver and banknote roubles. This fact is vividly described in Gogol's classic *Dead Souls*. When the landowner Sobakevich is bargaining furiously with the adventurist Chichikov over the dead souls, he asks "seventy-five roubles per soul, in banknotes as a favour to you". The same price in silver roubles would have been far more exorbitant. Although prices in banknotes were widespread, in fact they were only the price in silver doubled or tripled. The fact that here we are dealing with silver instead of gold makes no difference.

Anyone can see that the situation is now different. Metal money does not circulate at all, prices are not expressed in it, and gold itself has a real price in paper-credit money. The vast amount of the newly-produced gold is used in industry and bears no relation to the monetary sphere. Can we say that the function of gold as a measure of value has remained unchanged in these conditions? Evidently not.

Marx regarded money as the unity of its functions. He pointed to the conditions under which gold could serve fully as a universal equivalent of the commodity world: "Gold serves as an ideal measure of value, only because it has already, in the process of exchange, established itself as the money-commodity. Under the ideal measure of values there lurks the hard cash."[23] But these conditions have long been absent from the capitalist economy. If the hard cash does happen to fall into anyone's hands, it is not as money, but as a specific hoarding commodity.

The inexorable forces of historical development are driving gold out of the monetary sphere. But commodities must be compared in some way, their values contrasted and their prices expressed in something. For this purpose paper-credit money exists. There is little point in discussing whether this is a good or a bad thing, for it is the only possible reality.

The question of the present mechanism of the performance by money of the functions of a measure of value has occupied an important place in recent years in specialist literature in the USSR and other socialist countries. The most interesting works in this sphere belong to V. M. Usoskin. He writes: "How are we to approach the formulation of an integral and non-contradictory scientific conception of money in the economy of modern capitalism? The first and, perhaps, most important step towards elaborating such a conception should lie, to our mind, in discarding the obviously outmoded thesis that in modern conditions only gold can perform the role of a monetary commodity and accordingly of a universal equivalent in the internal circulation of capitalist countries."[24] Usoskin suggests the conception of modern paper-credit money as a *special monopolised commodity* which, in spite of the absence of intrinsic value, acquires value characteristics due to the fact that the state has a monopoly on the regulation and control of "production" and "supply" of money. Money is the object of an "artificial monopoly", which makes its supply forever limited and creates a scarcity factor. The following analogy may help us to understand the author's logic. If access to a theatrical entertainment was free and the auditorium (stadium, etc.) could accommodate all those who wished to see it, tickets for this entertainment would have no value. But in fact the capacity of the auditorium is always limited, and the management has a monopoly on the sale of tickets. Thanks to this the tickets acquire value. It is almost the same with money. If anybody could "produce" it, it would have no value. But the state (partly also the banking system, the author adds, and we must add too) has a monopoly on the "production" of money. Since money has a *utility,* i.e., the ability to be exchanged for commodities, it acquires a quasi-value. We refer the reader interested in this matter to the article in question and to other works on this subject.

Possibly paper-credit money as a monopolised commodity does really perform in a specific way the function of a measure of value. But this is obvious only if its supply is effectively limited, i.e., given the more or less normal functioning of the credit and monetary system and some acceptable rates of inflation. The question therefore arises: is there not a spontaneous return to gold as money in the event of a serious disruption of the system, in a state of "galloping" inflation? It is hard to answer this question for at least two reasons. First, we must initially agree on what we mean by "a return to gold". Secondly, no one has yet defined what inflation is still acceptable and what is a "serious disruption" of the credit and monetary system.

In a certain sense the importance of gold increases, of course, when there is strong inflation and paper-credit money is undermined. But experience shows that it increases *as a hoardable commodity*. Even in the case of hyper-inflation, when money sometimes depreciated millions of times over, it can hardly be said that there was a spontaneous return to gold as a medium of the measurement and comparison of values, that is, to gold as money. It is more likely that such local and temporary measures of value as salt, cigarettes or food coupons could arise.

All the same this must be qualified with a big BUT. It has already been said that the question of gold and money directly involves the important dialectical principle that a phenomenon never exists in its pure form, and that statements of the "yes-no" type cannot lead to the truth. The loss by gold of the functions of money is a long process and one can say that it is not yet completed.

Comparatively recently, in 1973, a state loan was issued in France, which was nicknamed "Giscard" on the stock exchange after the then Minister of Finance, Giscard d'Estaing. The bonds of this loan are cashed not at face value, but at a rate which is calculated with due account of changes in the Paris market price of gold. In the USA the ban on the gold clause is being abolished: juridically the sum of a future payment can be linked with the price of gold in paper dollars. It can be said that in these and similar cases gold acts indirectly as a measure of value or, to be more precise, a measure of the depreciation of paper-credit money. Of course, those who regard the demonetisation of gold as full and complete argue that one might as well attach the sum of paying off a loan or debt to the price of silver, zinc, oil and of any commodity or set of commodities. In fact, sets of commodities are sometimes used for this indexation. But as a single commodity, as far as we know, only gold is used in this function. In the vast ocean of commodities it still occupies a special place. This is largely a reflection of the role that gold still plays in the international economic sphere. After all, the central banks do not hold reserves of copper, tin or even silver, but stubbornly persist in holding gold reserves. In favour of the monetary life of gold is an age-old tradition, its illustrious past. It would be rash to discount this factor.

World money is the function that gold is keeping longest. This is perfectly natural: the replacement of gold by paper-credit money in a sphere where the jurisdiction of the national state does not operate is particularly difficult and in present conditions hardly possible in a final way.

Some people think that it is impossible to do without gold as a measure of the value of national currency units. This is not so. The gold parities of currencies have now been abolished, and this has not had a significant influence on the economy. Far more important is the role of gold as a potential and real international means of payment. It is the only form of reserve in respect of which no question of credibility and acceptability arises. All other forms are credit in the economic sense. Dollar reserves are the debts of the USA to other countries with all the related consequences: they depreciate with inflation and may under certain conditions be frozen by the US government and used by it as a means of exerting pressure. Although the reserve positions of countries in the IMF and the Special Drawing Rights rest on the collective credit of many states, they are also reliable only in a situation of international trust, in normal conditions. No one can guarantee that these conditions will not change, that a country will not in some situation or other need reserves, the value and liquidity of which do not depend on external factors, on someonc else's will. These requirements are met by gold, and so far there are no signs that anything else could successfully take its place in this respect.

It may be assumed that gold has not lost the role of a military and strategic reserve, not as a raw material, of course, but as a universal means of payment, a reserve of world money. The repeated rises in the price of gold have sharply increased the real market value of gold stocks. Their ability to settle balance of payments deficits has grown accordingly. The movement of gold between central reserves and the market is being resumed in new forms. Since some countries are selling it on the market and others buying it, eventually the metal is transferred from some states to others. And this is the traditional function which it has always performed on the world market.

The yellow metal, which as money has been declared a barbarous relic and "cast out of good society", handed over to jewellers and hoarders, stubbornly refuses to leave the arena of international monetary relations. This has been neatly expressed by a Western economist who wrote that when Keynes labelled gold as a barbarous relic he was forgetting that we live in barbarous times. People and countries do not trust one another, and they look to gold as an absolute form of reserves.

What does the future hold in store for gold? History does not repeat itself. It seems inconceivable that there will be a return to the gold standard in any of its traditional forms. But a certain restoration of the monetary or quasi-monetary functions of gold, particularly in the international sphere, is possible and is actually taking place. Such predictions can often be found where you least of all expect them. In 1976 a book appeared that was co-authored by Herman Kahn, Director of the Hudson Institute in the United States, one of the most well-known futurologists. In the economic section of the forecast for the development of the Western world over the next twenty years the authors predict "a great development of indexed contracts and financial instruments and perhaps a renewed role for gold in the international monetary system (the latter mainly as a store of value and perhaps as *numeraire*). The SDR (Special Drawing Rights) will also be developed as a world currency, but will almost certainly fail to achieve full acceptance."[25]

A certain turn towards recognising the possibilities of gold in the international monetary system took place in the West in 1978-1980. Facts and opinions in support of this conclusion have been quoted above. The traditional demands that have bored everyone to death of the South African politicians, the French "metallists" and the American economists of supply-side faith to restore some form of gold standard will continue to meet with a sceptical reception. But without the noise and the pomp, without official recognition, gold will play its role in the international monetary system as a special and fundamentally indispensable component of reserve assets.

The role of gold will be connected largely with the evolution of the monetary system as a whole. After the conclusion of the Bretton Woods agreements it was generally thought that a stable and lasting international monetary system had been set up. To a certain extent this was true. Today there is nothing of the kind. The present monetary system is regarded as temporary, transitional. It is not at all clear, however, what this transition is leading to, where and in what conditions such a transition can take place.

In 1967, when the dollar was still convertible into gold and the Bretton Woods system still operating, the American Professor Kindleberger published an interesting work entitled *The Politics of International Money and World Language*. In it he compared gold as world money with Latin and said that nostalgic dreams about reviving the monetary role of gold are as utopian as plans for reviving Latin as a cosmopolitan language. He likened Special Drawing Rights to Esperanto, noting the lack of success of attempts to put it into international circulation. What remains? Kindleberger made the following conclusion: just as English is spontaneously becoming the main international language, so in practice the American dollar is the best international money. Incidentally, the system of floating exchange rates proposed at that time by "academic economists" and the more or less equal participation of many currencies in international circulation he compared with the Tower of Babel.[26]

First and foremost, let us say that the "Tower of Babel" in the monetary sphere has become a fact. Exchange rates are floating, and other currencies, apart from the dollar, are being more widely used as international reserves. The dollar of today is not the one for which the professor prophesied the role of natural and ever-lasting international money. Yet this dollar, inflationary, inconvertible and depreciating, still serves as a substitute for international money.

In a world where there is a Tower of Babel, and the whole edifice has such a shaky foundation, gold will inevitably maintain its position.

Gold and Capitalism. In Lieu of a Conclusion

Throughout this book there have been cited interesting assessments of the role and prospects of gold, provided by scholars and politicians, industrialists and writers, bankers and philosophers. Let us quote one more such assessment. In a volume of the New Encyclopaedic Dictionary (a pre-revolutionary Russian encyclopaedia) published on the eve of World War I, the article on gold begins as follows: "The economic significance of gold is enormous. To say nothing of its technical use, the monetary systems of almost all the world rest on gold—now more than ever before. There is reason to believe that this function of gold, as also the size of its output, are approaching their peak." The author goes on to say: "Present-day monetary systems reveal . . . signs of transition to other forms of means of payment extending the circulation of paper notes that rests on the juridical force and the economically regulating activity of the state."[27]

In relation to the volume of gold output the prediction has not come true: between 1913 and 1970 it doubled in the capitalist world. As for the monetary functions of gold, they really did reach their peak by the beginning of World War I. In general the economic significance of gold has diminished compared with the beginning of the 20th century. The word "enormous" would hardly be appropriate to describe it today. This is, of course, explained by the fact that gold has ceased to be the basis of the monetary systems. Its role has changed not only quantitatively, but also qualitatively.

Let us try to test this thesis on the "hyperboloid model". I refer to the subject of Alexei Tolstoy's fantastic novel *Engineer Garin's Hyperboloid* mentioned in the Introduction. Let us imagine that Tolstoy's fantasy and the dreams of the charming and criminal adventurist Garin have come true: a way has been discovered of producing gold at a cost no higher, say, than that of producing copper. What would happen to the capitalist economic system then? Would there be an economic crisis and political collapse of the existing regime in the USA, as in *Hyperboloid*?

Frankly speaking, it would be hard to imagine such consequences resulting from a depreciation of gold, even if its price dropped thousands of times. There would probably be some dramatic events: crowds of people would besiege the places where gold was being sold for a few dollars a kilogram; the organised gold markets would close; the shares in gold-mining companies would plummet, and this could even cause general panic on the stock exchange. The central gold reserves would lose their value, hoarders would lose more or less, depending on how much of their capital was invested in gold. Then a more long-term readjustment would begin: the economy of the jewelry industry would change, the electronics and some other industries would begin to make more extensive use of gold...

But all this is very remote from economic and socio-political catastrophe, from the collapse of the system. The general (absolute) level of commodity prices and wage rates in dollars and other currency units would not change or, at least, would change slightly under the influence of indirect and remote effects of the depreciation of gold. Relative commodity prices would not move either, not counting the abrupt fall in the purchasing power of gold and, possibly, some kind of influence on the prices of substitute commodities, etc. Although gold would cease to play the role of monetary assets, there would probably not be any sudden catastrophic changes in the international monetary system either. In particular, the relationships between currencies, which play a decisive role today, would probably not change sharply under the influence of this factor as such.

Let us return from the realm of fantasy to reality. Our model illustrates only the well-known fact of the reduction in the monetary functions of gold under modern capitalism. But, as the reader can see, with regard to the economic life of gold as a whole the matter is rather complex. There has been and still is what one might call not so much a reduction as a change in the economic role of gold.

As already mentioned, the use value (economic utility) of any commodity is of a changing nature. This is perhaps even more true of gold than of other commodities: its utility is determined more by purely social factors. New forms of the use of gold are partly conditioned by the development of productive forces—methods of production, technology, and new types of products. Electronics is the most obvious example. On the other hand, socio-economic factors also have an influence—the evolution of capitalism as a social system, changes in the structure of society, in the role of the state and in social psychology. These factors have determined the decrease in the monetary functions of gold. But they also give rise to new fields of application, new forms of the economic life of gold. The first and second type of factors are closely intertwined. The growth of the use of gold in jewelry partly reflects the development of productive forces and changes in the structure of consumption. But to an even greater extent it is connected with the development of the structure of bourgeois society, the specific features of the bourgeois way of life. Gold serves ostentatious and prestigious consumption, hoarding, the concealing of incomes and wealth from the law and fiscal authorities. These features have long been inherent in capitalism and are only strengthened with its development. As we know, they do not disappear of their own accord even after the abolition of capitalism. They are the survivals in people's psychology and behaviour which socialism also encounters.

Among the factors determining the fate of gold a special role is played by inflation and socio-political tension. They stimulate hoarding, investment and speculative demand for gold and create a special and quickly growing field of application for it. There is no reason to believe that the significance and power of these factors will decrease.

Contradictory and complex tendencies are developing in the field of application of gold in the international monetary system. Gold cannot objectively be the basis of the modern monetary system. But whether this system can exist without gold is not yet proven. There can hardly be any doubt that the future evolution of the system will be linked with gold in some way or other.

The yellow devil can change his face. He hides behind the respectable façade of progress, credit, technology, and art. But as long as capitalism exists, the yellow devil lives on too. In people's minds he remains as before the symbol of bourgeois civilisation, the civilisation of money. And with good reason.

Gold has yet many surprises in store for mankind. The future of this social phenomenon will probably be no less rich in events and paradoxes than its past. If the reader of this book is better prepared to understand these events and paradoxes, the author will be able to consider his task fulfilled.

NOTES

I. INTRODUCTION. GOLD IN THE LIFE OF MANKIND

1. *Great Soviet Encyclopaedia,* third edition, Vol. 9, Sovietskaya entsyklopedia, Moscow, 1972, p. 567 (in Russian).
2. V. I. Lenin, "The Importance of Gold Now and After the Complete Victory of Socialism", *Collected Works,* Vol. 33, Progress Publishers, Moscow, p. 113.
3. Honoré de Balzac, *La Comédie humaine,* Vol. 2, Editions de la nouvelle revue française, Paris, 1935, p. 636.
4. Karl Marx, *Capital,* Vol. I, Progress Publishers, Moscow, 1974, p. 97.
5. S. M. Borisov, *Gold in the Economy of Modern Capitalism,* Finansy, Moscow, 1968, pp. 252, 254.
6. See Karl Marx, *A Contribution to the Critique of Political Economy,* Progress Publishers, Moscow, 1978, p. 64.
7. Thomas Hood, *Miss Kilmansegg and Her Precious Leg. A Golden Legend,* Samuel Buckley & Co., New York, 1904, p. 97.
8. A. N. Tolstoy, *Engineer Garin's Hyperboloid,* in *Complete Works,* Vol. 5, GIKhL, Moscow, 1947, p. 251 (in Russian).

II. HOW GOLD BECAME MONEY

1. Verney Lovett Cameron, *Across Africa,* Vol. 1, Bernhard Tauchnitz, Leipzig, 1877, pp. 211-212.
2. Carl Menger, *Gesammelte Werke,* Vol. I, *Grundsätze der Volkswirtschaftslehre,* J.C.B. Mohr, Tübingen, 1968, pp. 280-281.
3. *The History of Herodotus,* Vol. 1, E.P. Dutton and Co., New York, 1916, p. 50.
4. Ibid., p. 23.
5. *Currencies of the World. A Reference Book,* Edited by S. M. Borisov et al., Finansy, Moscow, 1976, p. 368 (in Russian).
6. See Robert Triffin, *The Evolution of the International Monetary System: Historical Reappraisal and Future Perspectives,* Princeton University Press, Princeton, 1964, p. 59.
7. See Arthur Nussbaum, *A History of the Dollar,* Columbia University Press, New York, 1957, p. 104.

III. MINING THE YELLOW METAL

1. Felix Krivin, *The Not Serious Archimedes,* Molodaya gvardia, Moscow, 1971, p. 60 (in Russian).

2. A. S. Pushkin, "The Covetous Knight", *Selected Works* in two volumes, Vol. 1, Progress Publishers, Moscow, 1976, p. 118.
3. See N. V. Petrovskaya, *Virgin Gold. A General Characterisation, Typomorphism, and Questions of Genesis,* Nauka, Moscow, 1973, p. 13 (in Russian).
4. See V. I. Sobolevsky, *Noble Metals. Gold,* Znanie, Moscow, 1970, p. 17 (in Russian).
5. Ibid., p. 40.
6. See Heinrich Quiring, *Geschichte des Goldes. Die goldenen Zeitalter in ihrer kulturellen und wirtschaftlichen Bedeutung,* F. Enke, Stuttgart, 1948, p. 217.
7. See N. V. Petrovskaya, op. cit., p. 237.
8. See O. L. Almazova, I. P. Bozhenko, I. G. Doronin, "The Production, Consumption and Markets of Gold in the Capitalist Countries", *Bulletin of Foreign Commercial Information,* Supplement 3, Market Research Institute, USSR Ministry of Foreign Trade, Moscow, 1974, p. 61 (in Russian).
9. *Best Short Stories of Jack London,* The Sun Dial Press, New York, 1945, p. 260.
10. Timothy Green, *The World of Gold Today,* Walker and Company, New York, 1973, pp. 53-54.
11. See O. L. Almazova et al., op. cit., p. 114.
12. Timothy Green, op. cit., p. 72.
13. See Erik Chanel, *L'Or,* Retz-C.E.P.L., Paris, 1974, p. 233.
14. See Franklyn Hobbs, *Gold, the Real Ruler of the World,* The Business Foundation Publishers, Chicago, 1943, p. 125.
15. See Heinrich Quiring, op. cit., p. 44.
16. Ibid., p. 123.
17. See Daniel Defoe, *Robinson Crusoe,* Prench, Trubner and Co., London, 1933.
18. See S. V. Potemkin, *The Noble 79th,* Nauka, Moscow, 1978, pp. 39-43 (in Russian).
19. See Adolf Soetbeer, *Edelmetallproduktion und Werthverhältnis zwischen Gold und Silber seit der Entdeckung Amerikas bis zur Gegenwart,* Justus Perthes, Gotha, 1879, pp. 107-111.
20. See Heinrich Quiring, op. cit., pp. 286-287.
21. See Adolf Soetbeer, op. cit., p. 110.
22. Ibid., p. 39.
23. Timothy Green, op. cit., pp. 30-31.
24. See Stefan Zweig, *Sternstunden der Menschheit. Zwölf Historische Miniaturen,* Aufbau-Verlag, Berlin, 1974, pp. 141-152.
25. See Eduard Suess, *Die Zukunft des Goldes,* Wilhelm Braumüller, Wien, 1877.
26. Paul Einzig, *The Future of Gold,* The Macmillan Company, New York, 1935, pp. 67, 63.
27. See Heinrich Quiring, op. cit., pp. 34, 112, 113.
28. Ibid., p. 156.

29. "Engels an Karl Kautsky in Wien, 5 Sept. 1889", in Marx/Engels, *Werke,* Vol. 37, Dietz Verlag, Berlin 1967, p. 274.
30. "Engels to C. Schmidt, October 27, 1890", in Karl Marx and Frederick Engels, *Selected Works* in three volumes, Vol. 3, Progress Publishers, Moscow, 1976, p. 490.
31. See D. N. Mamin-Sibiriak, *Gold,* in *Collected Works* in eight volumes, Vol. 6, Khudozhestvennaya Literatura, Moscow, 1955, pp. 74, 78 (in Russian).
32. See G. V. Foss, *Gold. Types of Deposits, History of Mining, and Raw Material Bases,* Gosgeoltekhizdat, Moscow, 1963, pp. 119-120 (in Russian).
33. See Allan Patrick Cartwright, *The Gold Miners,* Purnell and Sons, Cape Town, Johannesburg, 1963, p. 85.
34. See Timothy Green, op. cit., p. 67.
35. Ibid., p. 60.

IV. WHERE THE GOLD GOES

1. See S. M. Borisov, *Gold in the Economy of Modern Capitalism,* Finansy, Moscow, 1968, p. 9 (in Russian).
2. See A. Kladt, V. Kondratiev, *The Story of the "Gold Train",* Politizdat, Moscow, 1966, p. 100 (in Russian).
3. See Timothy Green, *The World of Gold Today,* Walker and Company, New York, 1973, p. 127.
4. Timothy Green, op. cit., p. 173.
5. See *The New York Times,* September 24, 1974, pp. 57, 65; *New York Herald Tribune,* September 24, 1974.
6. René Sédillot, *Les secrets du marché de l'or,* Recueil Sirey, Bordeaux, 1948, pp. 36-37.
7. Gunnar Myrdal, *An International Economy,* Routledge & Kegan Paul Ltd., London, 1956, p. 364.
8. See *The Banker,* Vol. 125, No. 589, March 1975, p. 261.
9. See Timothy Green, op. cit., p. 134.
10. Estimates based on data published in the sources: *The Banker,* Vol. 125, No. 589, March 1975, p. 261; *Gold 1977,* Consolidated Gold Fields, London, 1977.
11. Timothy Green, op. cit., p. 250.
12. See René Sédillot, *Histoire de l'or,* Fayard, Paris, 1972, p. 351.
13. See Timothy Green, op. cit., p. 250.
14. See *Gold 1977,* Consolidated Gold Fields London, 1977.
15. See René Sédillot, *Les secrets du marché de l'or,* p. 49.
16. See Timothy Green, op. cit., p. 263.
17. Ibid., p. 253.
18. Charles Montesquieu, *De l'esprit des lois,* Editions sociales, Paris, 1969, p. 159.
19. See Arthur Hailey, *The Moneychangers,* Doubleday & Company Inc., New York, 1975, p. 304.
20. Timothy Green, op. cit., p. 192.
21. See *Gold 1979,* Consolidated Gold Fields, London, 1979, p. 54.

22. See *Annual Bullion Review 1978,* S. Montagu and Co., London, March 1979, p. 22.
23. *The Economist,* Vol. 274, No. 7115, January 12, 1980, p. 85.
24. See Timothy Green, op. cit., p. 238.
25. René Sédillot, *Histoire de l'or*, p. 353.
26. See *Gold 1977,* Consolidated Gold Fields, London, 1977, p. 22.
27. See S. M. Borisov, op. cit., p. 251.
28. See *Gold 1979,* Consolidated Gold Fields, London, 1979, pp. 18, 22.
29. See Timothy Green, op. cit., pp. 213-214.
30. Ibid., p. 215.
31. See René Sédillot, op. cit., p. 339.
32. See Timothy Green, op cit., p. 24.
33. Ibid., p. 222
34. René Sédillot, op. cit., p. 337.
35. Karl Marx, *Capital,* Vol. I, Progress Publishers, Moscow, 1974, p. 93.

V. GOLD GENOCIDE

1. A. S. Pushkin, *Collected Works* in ten volumes, Vol. 2, Khudozhestvennaya literatura, Moscow, 1974, p. 96 (in Russian).
2. *The History of Herodotus,* Vol. 1, E. P. Dutton and Co., New York, 1916, pp. 95, 96.
3. Heinrich Quiring, *Geschichte des Goldes. Die goldenen Zeitalter in ihrer kulturellen und wirtschaftlichen Bedeutung,* F. Enke, Stuttgart, 1948, p. 99.
4. Suetone, *Les douze Césars,* Vol. 1, Librairie Garnier Frères, Paris, 1931, pp. 58-59.
5. Karl Marx, *Capital,* Vol. I, Progress Publishers, Moscow, 1974, p.226.
6. See Heinrich Quiring, op. cit., p. 124.
7. See Bartolomé de las Casas, *Historia de las Indias,* Vol. 2, Editora Nacional, Mexico, 1877, pp. 200-201.
8. Christopher Columbus, *Four Voyages to the New World. Letters and Selected Documents,* Corinth Books, New York, 1962, p. 12.
9. Ibid., p. 196.
10. See Bartolomé de las Casas, op. cit., p. 9.
11. See Heinrich Quiring, op. cit., p. 206.
12. See Adolf Soetbeer, *Edelmetallproduktion und Werthverhältnis zwischen Gold und Silber seit der Entdeckung Amerikas bis zur Gegenwart,* Justus Perthes, Gotha, 1879, p. 107.
13. *Vision de los vencidos,* Casa de las Américas, Havana, 1969, p. 70.
14. See Heinrich Quiring, op. cit., pp. 219-220.
15. See Jean-Antoine Llorente, *Histoire critique de l'inquisition d'Espagne,* Vol. 1, Chez Treuttel et Würtz, Paris, 1818, p. 262.
16. See Hans Roden, *Treasure-Seekers. A Chronicle of Fortunes Lost and Hidden and the Efforts Made to Salvage Them,* George G. Harrap & Co. Ltd., London, 1966, pp. 66-67.
17. See Robert Louis Stevenson, *Treasure Island,* T. Nelson & Sons, London, s.d., p. 273.

18. See *All the Monies of the World. A Chronicle of Currency Values,* Frank Pick, René Sédillot, ed. Pick Publishing Corporation, New York, 1971, p. 130.
19. Alexander Del Mar, *A History of the Precious Metals from the Earliest Times to the Present,* George Bell and Sons, London, 1880, pp. 310-311.
20. See *Gold Rush Album.* Edited by Joseph Henry Jackson, Charles Scribner's Sons, New York, 1949, p. 22.
21. William P. Morrell, *The Gold Rushes,* A. and Ch. Black, London, 1940, p. 190.
22. See *SShA—ekonomika, politika, ideologia,* No. 8, Moscow, 1977, pp. 126-127.
23. F. I Mikhalevsky, *Gold in the Period of the World Wars,* Politizdat, Moscow, 1945, p. 184 (in Russian).
24. Hans Roden, op. cit., pp. 180-194.
25. See Francis Wilson, *Labour in the South African Gold Mines. 1911-1969,* Cambridge University Press, Cambridge, 1972, p. 21.
26. Timothy Green, *The World of Gold Today,* Walker and Company, New York, 1973, p. 78.
27 Kenneth Blakemore, *The Book of Gold,* Stein and Day Publishers, New York, 1971, p. 50.
28. *Albrecht Dürers schriftlicher Nachlass,* Verlegt bei Julius Berd, Berlin, 1920, pp. 47, 48.

VI. THE POWER OF GOLD

1. Maxim Gorky, *The City of the Yellow Devil,* Progress Publishers, Moscow, 1977, pp. 17, 18.
2. Franklyn Hobbs, *Gold, the Real Ruler of the World,* The Business Foundation Publishers, Chicago, 1943, pp. 96, 181 et al.
3. John K. Galbraith, *Money. Whence It Came, Where It Went,* Penguin Books, London, 1976, p. 109.
4. For more detail see: A. V. Anikin, *A Science in Its Youth,* Progress Publishers, Moscow, 1975, Chapter 5.
5. A. S. Pushkin, *Collected Works* in ten volumes, Vol. 4, Khudozhestvennaya literatura, Moscow, 1975, p. 353 (in Russian).
6. Karl Marx, *Capital,* Vol. I, Progress Publishers, Moscow, 1974, pp. 131-132.
7. René Sédillot, *Histoire de l'or,* Fayard, Paris, 1972, pp. 7, 10.
8. *The Holy Bible,* Cambridge University Press, London, 1903, p. 2.
9. See Heinrich Quiring, *Geschichte des Goldes...,* p. 26.
10. *The Holy Bible,* p. II.
11. Karl Marx, *Capital,* Vol. I, p. 96.
12. Franklyn Hobbs, op. cit., p. 211.
13. Ibid., p. 192.
14. René Sédillot, op. cit., p. 2.
15. Paul Einzig, *The Destiny of Gold,* The Macmillan Press Ltd., London, 1972, p. 14.
16. Karl Marx, *Capital,* Vol. I, p. 132.
17. Paul Einzig, op. cit., p. 6.

18. Felix Krivin, *The Not Serious Archimedes,* Molodaya gvardia, Moscow, 1971, p. 149 (in Russian).
19. Bernard Shaw, *The Intelligent Woman's Guide to Socialism, Capitalism, Sovietism and Fascism,* Constable and Company Ltd., London, 1949, p. 263.
20. Franklyn Hobbs, op. cit., p. 227.
21. See Z. S. Katsenelenbaum, *South African Gold and the Aggravation of Anglo-American Contradictions,* Gosfinizdat, Moscow, 1954, p. 95 (in Russian).
22. See M. A. Barg, *Shakespeare and History,* Nauka, Moscow, 1976, p. 162 (in Russian).
23. William Shakespeare, *Timon of Athens,* Arden Shakespeare Paperbacks, Methuen & Co. Ltd., London, 1977, pp. 91-92.
24. *Economistes financiers du XVIII-e siècle,* Chez Guillame Libraire, Paris, 1843, pp. 394-395.
25. A. N. Radishchev, *A Journey from St Petersburg to Moscow, Complete Works,* Vol. 1, USSR Academy of Sciences Publishing House, Moscow-Leningrad, 1938, p. 383 (in Russian).
26. John M. Keynes, *A Tract on Monetary Reform,* Macmillan and Co. Ltd., London, 1924, p. 172.
27. Arthur Hailey, *The Moneychangers,* Doubleday & Company Inc., New York, 1975, p. 136.
28. See *Pick's Currency Yearbook 1973,* Pick Publishing Corporation, New York, 1973, p. 593.
29. See Allan Patrick Cartwright, *The Gold Miners,* Purnell and Sons, Cape Town, Johannesburg, 1963, p. 82.
30. Eric Chanel, *L'or,* Retz-C.E.P.L., Paris, 1974, p. 233.
31. Allan Patrick Cartwright, op, cit., p. 194.
32. Eric Chanel, op. cit., p. 15.

VII. THE GOLD STANDARD AND ITS AGONY

1. Adam Smith, *An Inquiry into the Nature and Causes of the Wealth of Nations,* Vol. 1, Oxford University Press, London, 1928, p. 32.
2. John K. Galbraith, *Money. Whence It Came, Where It Went,* Penguin Books, London, 1976, p. 51.
3. Statistical data from the books: S. M. Borisov, *Gold in the Economy of Modern Capitalism,* Finansy, Moscow, 1968, pp. 9, 92, 107, 112 (in Russian); Robert Triffin, *The Evolution of the International Monetary System: Historical Reappraisal and Future Perspectives,* Princeton University Press, Princeton, 1964, p. 57.
4. Karl Marx, *Capital,* Vol. III, Progress Publishers, Moscow, 1977, pp. 572, 573.
5. Ibid., p. 606.
6. John K. Galbraith, op. cit. p. 178.
7. Karl Marx, *Capital,* Vol. III, p. 490.
8. See S. M. Borisov, op. cit., p. 253.
9. See F. I. Mikhalevsky, *Gold in the Period of the World Wars,* Politizdat, Moscow, 1945, p. 140 (in Russian).
10. See John K. Galbraith, op. cit., p. 271.

11. See S. M. Borisov, op. cit.; IMF Statistics.
12. See S. M. Borisov, op. cit., p. 184.
13. *Le Monde,* 6 February 1965.

VIII. THE INTERNATIONAL MONETARY SYSTEM

1. John K. Galbraith, *Money. Whence It Came, Where It Went,* Penguin Books, London, 1976, p. 312.
2. *The Banker,* Vol. 125, No. 593, July 1975, p. 793.
3. See Robert L. Heilbroner, *Beyond Boom and Crash,* W. W. Norton and Co. Inc., New York, 1978.
4. See *International Financial Statistics, July 1979,* IMF, Washington, 1979.
5. *The Economist,* Vol. 274, No. 7125, March 22, 1980, p. 10.
6. *Newsweek,* Vol. LXXVIII, No. 9, August 30, 1971, p. 12.
7. Robert Solomon, *The International Monetary System, 1945-1976. An Insider's View,* Harper & Row Publishers, New York, 1977, p. 180.
8. D. V. Smyslov, *The Crisis of the Modern Capitalist Monetary System and Bourgeois Political Economy,* Nauka, Moscow, 1979, p. 353 (in Russian).
9. Robert Solomon, op. cit., p. 205.
10. See *International Monetary Fund. Annual Report 1979,* Washington, 1979, p. 41.
11. See Paul Einzig, *The Destiny of Gold,* The Macmillan Press Ltd., London, 1972, p. 1.
12. See Robert Solomon, op. cit., p. 235.
13. See *International Financial Statistics, May 1977,* IMF, Washington, 1977.
14. *The Economist,* Vol. 274, No. 7125, March 22, 1980, p. 28.
15. V. I. Lenin, "Preface to N. Bukharin's Pamphlet, *Imperialism and the World Economy", Collected Works,* Vol. 22, p. 107.
16. See *International Financial Statistics, March 1978,* IMF, Washington, 1978.
17. *International Monetary Fund. Annual Report 1979,* Washington, 1979, p. 67.
18. J. Marcus Fleming, *Reflections on the International Monetary Reform,* Princeton University Press, Princeton, 1974, p. 1.
19. D. V. Smyslov, op. cit., Chapters 13, 14.
20. Robert Solomon, op. cit., p. 315.
21. Ibid., p. 317.
22. *International Monetary Fund. Annual Report 1979,* pp. 78-81.
23. *The Economist,* Vol. 274, No. 7125, March 22, 1980, p. 81.
24. *Le Figaro,* April 21, 1978, p. 7.

IX. THE FATE OF THE YELLOW METAL

1. I. A. Trakhtenberg, *Money and Credit under Capitalism,* USSR Academy of Sciences Publishing House, Moscow, 1962, p. 27 (in Russian).

2. Robert Triffin, "The Demonetisation of Gold: Pros and Cons", *Economic Impact,* No. 10, Washington, 1975, p. 51.
3. S. M. Borisov, *Gold in the Economy of Modern Capitalism,* Finansy, Moscow, 1968, p. 160.
4. Pierre Vilar, *Or et monnaie dans l'histoire 1450-1920,* Flammarion, Paris, 1974, p. 97.
5. Roy W. Jastram, *The Golden Constant. The English and American Experience, 1560-1976,* John Wiley & Sons, New York, 1977, p. 34.
6. Robert Triffin, *The Evolution of the International Monetary System: Historical Reappraisal and Future Perspectives,* Princeton University Press, Princeton, 1964, p. 13.
7. John K. Galbraith, *Money. Whence It Came, Where It Went,* Penguin Books, London, 1976, p. 224.
8. Roy W. Jastram, op. cit., pp. 147-148.
9. *The Economist,* Vol. 248, No. 6779, July 28, 1973, p. 62.
10. *The Future of the World Economy. A Study on the Impact of Prospective Economic Issues and Policies on the International Development Strategy,* United Nations, New York, 1977, p. 65.
11. Roy W. Jastram, op. cit., p. 126.
12. Ibid., pp. 33-36.
13. Ibid., pp. 145-146.
14. Ibid., pp. 26-33.
15. Ibid., p. 188.
16. *The Future of the World Economy...*, pp. 65-66.
17. *The Economist,* Vol. 274, No. 7115, January 12, 1980, p. 86.
18. *The Economist,* Vol. 274, No. 7125, March 22, 1980, p. 81.
19. *Money and Credit in the Capitalist Countries,* Edited by L. N. Krasavina, Finansy, Moscow, 1977, p. 20 (in Russian).
20. L. N. Krasavina, "Crisis Phenomena in the Monetary Sphere of the West", *International Affairs,* No. 1, Moscow, 1980, p. 42.
21. *Journal officiel de la République Française. Débats parlémentaires. Assemblée nationale. Séance du 17 Octobre 1963,* Paris, 1963, pp. 5219-5220.
22. *The Journal of Commerce,* New York, January 24, 1980, pp. 1, 3.
23. Karl Marx, *Capital,* Vol. I, Progress Publishers, Moscow, 1974, pp. 105-106.
24. V. M. Usoskin, "The Money of Monopoly Capitalism", *Mirovaya ekonomika i mezhdunarodniye otnosheniya,* No. 3, 1979, p. 54.
25. Herman Kahn, William Brown and Leon Martel, *The Next 200 Years. A Scenario for America and the World,* Associated Business Programmes, London, 1977, pp. 191-192.
26. Charles P. Kindleberger, *The Politics of International Money and World Language,* Princeton University Press, Princeton, 1967, pp. 10-11.
27. *The New Encyclopaedic Dictionary,* F. A. Brockhauz and I. A. Efron, Vol. 18, St Petersburg, 1914, p. 795 (in Russian).